ALCOHOL and CAFFEINE

> A Study of Their Psychological Effects

By

HARVEY NASH, Ph.D.
Northwestern University Medical School
Chicago, Illinois

CHARLES C THOMAS • PUBLISHER
Springfield • Illinois • U.S.A.

CHARLES C THOMAS • PUBLISHER
BANNERSTONE HOUSE
301-327 East Lawrence Avenue, Springfield, Illinois, U.S.A.

This book is protected by copyright. No
part of it may be reproduced in any manner
without written permission from the publisher.

© *1962, by* CHARLES C THOMAS • PUBLISHER
Library of Congress Catalog Card Number: 61-17615

*With THOMAS BOOKS careful attention is given to all details of
manufacturing and design. It is the Publisher's desire to present books
that are satisfactory as to their physical qualities and artistic possibilities
and appropriate for their particular use. THOMAS BOOKS will be true
to those laws of quality that assure a good name and good will.*

Printed in the United States of America

TO THE READER

REPORTED HERE is an experimental study of alcohol and caffeine effects. The results of the experiment are evaluated in terms of, and co-ordinated with, the findings of previous investigators.

This was a large experiment, in more than one respect. It included a sizable number of psychological tests, representing many and varied psychological functions. And though the number of subjects participating in the experiment was not large, it exceeded the numbers examined in the great majority of alcohol and caffeine studies conducted previously.

A decided effort was made to identify and control possible sources of error while the investigation was being planned and conducted, and to evaluate the relative contributions of whatever sources of error did manage to influence the results. The introduction of a variety of procedures designed to eliminate or at least to regulate error made for a complex experiment. Readers insufficiently acquainted with the psychological, pharmacological, and statistical aspects of psychopharmacological research might gain a fuller understanding of the experiment's technical features by examining the author's "The design and conduct of experiments on the psychological effects of drugs" (see References).

Because of the care that went into its design and execution, and because of its magnitude and scope, the present experiment provides a suitable framework around which to integrate the earlier experimental work on alcohol and caffeine. This report is not, however, meant to include a complete review of the literature; the present discussion considers only those earlier findings pertinent to the functions investigated here.

The sections of the report follow one another in a logical order, but readers may wish to take some liberties with the order in which sections are read. Thus, Parts III and IV as well as Appendix A—these sections develop certain general implications of the findings—might well be perused after the first two chapters have been read. Reading might then be resumed with Chapter 4. Chapter 3—which describes the test procedures—can be treated as a reference section and skipped over; it can then be referred to when particular tests are discussed in later chapters. By organizing his reading along lines such as these, even the reader having little familiarity with studies of drug effects on psychological functions may derive appreciable benefit from the report.

ACKNOWLEDGMENTS

THIS INVESTIGATION was supported by a grant from the Mental Health Fund of the Illinois Department of Public Welfare. I am indebted to Professor George Yacorzynski and to the members of the Illinois Psychiatric Research and Training Authority, especially Professor Benjamin Boshes, for their encouragement and support. Gertrud Schaffner's efforts on the project's behalf were varied and tireless. Extensive efforts were also put forth by Professor A. E. Zeller, who explored methods for determining drug concentrations in small samples of blood, and by Judith Keeler, Alice Taylor, and Edna Cunningham. Scoring services were provided by P. R. Merrifield and by Janet Hirsh. The illustrations were prepared by Edith Hodgson. I wish to express my gratitude to these individuals.

I am further grateful to Professor K. A. Brownlee for his valuable assistance in the analysis of results; to Professors Brownlee, Joseph E. Barmack, and J. P. Guilford, and to Dr. Hannah Steinberg for their critical readings of early drafts of the manuscript; and to Jean Ratliff for editorial assistance.

I wish to thank Professor Guilford, for permission to use tests prepared under U. S. Government Contract N6onr-23810; Professor Irving J. Saltzman, for permission to adapt a test of incidental learning; Professor Morris I. Stein, for permission to adapt Stein's Sentence Completion Test; Professor Thelma G. Thurstone, for permission to use or adapt tests developed by the late Professor L. L. Thurstone; Science Research Associates, for permission to adapt Cardall's Test of Practical Judgment; and the Psychological Corporation, for permission to adapt a number of their published tests.

Lastly, I would like to express my appreciation to the subjects of the study, whose co-operative attitude increased the likelihood of valid findings.

HARVEY NASH

CONTENTS

	Page
To the Reader	v
Acknowledgments	vii

PART I
The Problem and the Method

Chapter
1. Introduction 5
2. Design of the Study 9
   Subjects 9
   Experimental sessions 10
   Experimental treatments 12
   Blood samples 15
3. Psychological Procedures 16
   Sessions 1 and 3 17
   Sessions 2 and 4 21
   Self-ratings 28
   Reaction to puncture 28

PART II
Results and Discussion

4. Background Information 33
   Subjects 33
   Blood levels for alcohol and caffeine 35
   Analysis of data and tabular presentation of results . . 36
5. Self-Rating of Subjective States 40

Chapter	Page
6. Intellectual Functioning	45
Capacity measures, over-all findings	45
Motor control, capacity measures	48
Motor control, incidental measures	50
Associative productivity	56
"Quality" of associations	61
Fixed associations, overlearned	61
Visual thinking	63
Visual-motor co-ordination	68
Recall	70
Learning	75
Reasoning	76
Flexibility of thinking	77
Critical judgment	81
Discrimination of significant features in practical social situations	84
7. Affectivity and Response to Stress	86
Affectivity	86
Reaction to bodily threat	86
Resolution of competing tendencies	89
Reaction to stress (projective)	90
Summary of results for stress reactions	92
8. Regression on Blood Levels and on Customary Consumption of Drugs	95

PART III
General Comments

9. Further Discussion of Present Findings	103
Considerations affecting evaluation of over-all drug effects	103
a. Two-tailed tests	103
b. Confounding due to "side effects" of the heavy alcohol dose	103
c. Suggestion	105
d. Practice effects and confounding	105
Barmack's hypothesis	106
Practical versus statistical significance	107

Contents

Chapter	Page
10. Alcohol Levels and Direction of Drug Effects	110
Drug effects as a function of alcohol level	110
Biphasic alcohol effects and the disinhibition hypothesis	112
11. Task Features Affecting Response to Drugs	115
Task complexity	115
Implicit versus explicit tasks	117
Task familiarity	117
The role of attention	118
Test length	119
Tasks requiring rapid functioning	121

PART IV

Conclusion

12. Summary and Conclusions	125

Appendices

A. The Double-Blind Procedure: Rationale and Empirical Evaluation	133
Problem	133
Suggestion	134
Features of the double-blind procedure	135
An empirical evaluation of double-blind procedures	139
1. Procedures of the present study	139
2. Evaluation of examiner blindness	141
3. Evaluation of subject blindness	143
Conclusion	146
B. The Arithmetic Errors Test	149
C. Clusters for the Sentence Completion Test	150
References	151
Author Index	159
Subject Index	163

ALCOHOL AND CAFFEINE

PART I
The Problem and the Method

1 INTRODUCTION

DURING THE PAST decade we have witnessed a vast expansion in the number and variety of chemical agents that act on the central nervous system to alter psychological functioning. To our previous stock of drugs have been added such newly explored or newly developed psychochemicals as the energizers, the hallucinogens, and of greatest practical significance, the various kinds of tranquilizers. Research on the psychological effects of these drugs, motivated almost exclusively by psychiatric considerations and conducted largely with psychiatric patients or with animals, has increased at a corresponding pace. This heightened research activity has led research workers to distinguish a new crossroads of science, psychopharmacology.

Yet psychopharmacology is by no means a fresh scientific discipline. Some seventy years ago Kraepelin (1892) was already undertaking intensive investigations of the psychological effects of central nervous system depressants and stimulants. Although Kraepelin is best known for his later clinical contributions to psychopathology, his experimental researches, conducted in Wundt's laboratory, were primarily concerned with drug effects on normal human functioning.

Ethyl alcohol was perhaps foremost among the depressant drugs that claimed the attention of Kraepelin and his successors. Experiments on ethyl alcohol's effects have been motivated not only by interest in pure research but also by interest in such applied problems as alcoholism, temperance and prohibition, industrial productivity, and traffic safety, with the alcohol experiments of any given period tending to reflect the social concerns of that period. Despite the volume of accumulated experimental literature, however, our knowledge of alcohol's psychological effects remains surprisingly incomplete—due both to methodological shortcomings of the experimental literature and to difficulties that beset research on alcohol effects in human subjects.

Much attention has also been paid to central nervous system stimulants. Hollingworth (1912) reported an intensive investigation of caffeine effects some years after Kraepelin first worked with this agent. The development of the amphetamine compounds resulted in experimentation, first with racemic amphetamine (benzedrine) in the 1930's and 1940's, then with dextro-amphetamine (dexedrine) and methamphetamine (pervitin), especially during World War II; much of this research was inspired by interest in the fatigue-alleviating properties of the amphetamines.

These researches on "classical" depressant and stimulant drugs provide the present investigation's point of departure. In moderate quantities at least, many of these classical agents act primarily on the central nervous system's higher levels, modifying those aspects of behavior and experience that are most characteristically human. More than merely providing information on various psychological effects of drugs, psychological experimentation with classical psychochemical agents provides opportunities for relating alterations of "higher mental functions" to each other and to the central nervous "sub-systems" that mediate the psychological actions of these drugs.

The present investigation is concerned with the psychological effects on normal man of the classical psychochemicals alcohol and caffeine. Each drug affects a variety of psychological functions, and these drug actions are presumably effected over a range of c.n.s. sites rather that at a single site. For alcohol, at least, higher c.n.s. levels and higher mental functions tend to be affected most readily by mild doses; lower c.n.s. levels and more primitive psychological functions come to be affected with increasing dosage.

To help integrate the wealth of existing information about alcohol's psychological effects and about its dose-response characteristics, it was thought desirable to obtain dose-response data for alcohol. Alcohol was therefore administered in two strengths, to bring out this drug's typical psychological effects and to determine whether psychological functioning is related to alcohol dosage level in a simple linear fashion. Barely detectable effects on performances mediated by higher brain centers were expected from the smaller alcohol dose. Marked impairment of performances mediated by higher brain centers, and only a mild disturbance of lower centers, were expected from the larger (double-strength) alcohol dose.

Because caffeine has been studied much less extensively than alcohol, it was thought best to concentrate on helping firmly establish caffeine's *characteristic* psychological actions, and not diffuse the investigator's resources by also studying caffeine's dose-response characteristics. Caffeine was therefore administered in but a single strength, at a dosage level expected to produce near-optimal psychological effects. This investigation is thus asymmetrical, fewer resources having been allocated for studying caffeine effects than for studying alcohol effects.

Major questions posed by the investigation include:

1. Does alcohol impair, and caffeine enhance, performance on a majority of the psychological functions studied? Are the *over-all* actions of these two drugs opposite in direction?

2. Do alcohol and caffeine have *selective* actions on particular psychological functions, apart from whatever over-all actions they may manifest? Assuming that the actions of alcohol and caffeine are *generally* opposite in kind, are these actions also diametrically different for *each* individual function, or is there an *interaction* between drugs and functions? Whatever each drug's *predominant* action, depression or stimulation, do the alcohol effects form a pattern similar to or distinct from the pattern of caffeine effects?

3. Do the actions produced by different alcohol dosage levels differ in *kind* as well as in *degree*?

To provide answers to these questions, drug effects were compared on a variety of psychological functions represented by a large number of test performances. Special attention was focused on drug-induced changes in the subjects' capacities to cope with problems posed by a challenging environment. Studied to a lesser extent were drug-induced changes in subjective feelings, in perceptions of self and environment, and in characteristic modes of expression. Psychological tests were selected not only according to their content, but also on the basis of their reliability and brevity, and of the availability of equivalent test forms. Some tests of low reliability and even of uncertain validity were employed, however, to provide at least exploratory coverage for several important functions lacking adequate measuring instruments.

2 DESIGN OF THE STUDY

A TOTAL OF 56 subjects, each of whom received one of four doses, was examined before and after treatment. The present chapter indicates how Ss were selected and assigned to treatments. Also described here are the structure of the experimental sessions, the composition and administration of the doses, and the collection and analysis of blood samples. The psychological tests are described in Chapter 3, and the procedures intended to establish and maintain double-blind conditions are discussed in Appendix A.

Subjects: Openings for paid Ss, in a "medical study," were advertised in several college newspapers. Inquiring for further details, prospective applicants learned only that Ss drank alcoholic beverages, took psychological tests, donated very small quantities of blood, and did without a couple of meals—for which services they received $1.25 per hour plus carfare. Neither at this time nor at any later time was mention made of caffeine.

Applicants between the ages of 18 and 45, having at least some education beyond high school, were eligible to participate in the study. Excluded from the study were applicants:

1. Reporting physical conditions, such as peptic ulcer, that could be aggravated by the drug treatments, or that could interfere with the metabolism of ethyl alcohol or caffeine;

2. With a history of vomiting, marked headache, or similar physical upset following ingestion of alcohol;

3. With a history of marked emotional upset at the sight of blood, or on being punctured for blood-taking purposes;

4. Excessively sophisticated with regard to psychological testing procedures or with regard to the use of placebos in drug studies.

Fifty-six applicants, including thirteen women, were selected as Ss, and devoted an average of 13 hours each to the study.

Experimental Sessions: A number of psychological tests and rating scales were administered to Ss over a period of four sessions. Psychological procedures were administered under *drug-free* conditions during the *first two* sessions (because of the large number of tests, the entire group of tests was partitioned into two portions, and each portion was administered during a distinct session). These procedures were repeated with some modification during the *last two* sessions, *under the influence of the experimental treatments*.

Psychological tests for each session were grouped into small test batteries. The order of administration of batteries was constant during each of the first two (pre-treatment) sessions, and randomized from S to S (cf. Nash, 1959) during each of the final two (treatment) sessions; see Table 1. Tests *within* each battery were always presented in a fixed order, designed to minimize undesirable interactions between tests. (See Chapter 3.)

Ss were examined individually during treatment sessions to help maintain their "blindness" about the drinks (see Appendix A for a description of the double-blind procedure followed in this study), and to avoid interactions among drugged Ss (cf. Nowlis and Nowlis, 1956). Tests were administered to groups of one to four Ss during pre-treatment sessions; each group was restricted to members of a single "block" (see below). Group administration during pre-treatment sessions was made possible by the fixed order of test administration during these sessions. Face-to-face contact between the examiner and Ss examined individually was minimized, and direct or indirect communication (of attitudes or of specific information) among Ss examined in a group was discouraged, in order to minimize discrepancies in testing conditions between pre-treatment and treatment sessions.

The same examiner conducted all sessions and examined all

TABLE 1

Order of Events During the Four Sessions of the Experiment

First and Second Sessions (Pre-treatment)	Third and Fourth Sessions (Treatment)

TIME

2nd session only: Test battery F; Height, weight obtained; Vocabulary test

Time	Event
0[a]	
10	Initial dose, divided
20	
28	Three blood samples;
35	ratings of Reaction
42	to Puncture

Self-rating: Dislike Puncture
Ten Self-ratings ⟷ 45 Ten Self-ratings
Writing Speed (*1st session only*) ⟷ 48 Writing Speed (*3rd session only*)
50 Administration of test batteries begun

1st session only: Test batteries A through C (administered in consecutive order)

Testing interrupted for ten Self-ratings (omitted from analysis of results)

2nd session only: Test batteries G and H (administered in consecutive order)

Time	Event
M−5	Two blood samples; ratings of Reaction to Puncture
M−3	Ten Self-ratings (omitted from analysis of results)
M+0[a]	Maintenance dose administered
M+13	Three blood samples;
M+18	ratings of Reaction to
M+24	Puncture
M+27	Ten Self-ratings (omitted from analysis of results)

Testing interrupted for maintenance dose. Test batteries F through J (*4th session*, administered in random order). Test batteries A through E (*3rd session*, administered in random order).

Incidental Learning and other "post-battery" procedures (*4th session only*)

1st session only: Test batteries D and E (administered in consecutive order)

Writing Speed (*1st session only*) ⟷ Writing Speed (*3rd session only*)
Two blood samples; ratings of Reaction to Puncture
Ten Self-ratings ⟷ Ten Self-ratings
Self-rating: Dislike Puncture
Examiner's guess regarding treatment assigned S

---

[a] Time (in minutes) is indicated for procedures administered at predetermined times during treatment sessions. During the early parts of sessions 3 and 4, procedures were administered at specified intervals after "zero time" (the time at which S began to ingest the first portion of the initial drink). The reference point for procedures administered at predetermined times during the middle and late parts of sessions 3 and 4 was "M + 0" (the time at which S began to ingest the maintenance drink). Identical procedures administered at corresponding times during pre-treatment and treatment sessions are linked by arrows.

Ss. The examiner strove for objectivity in his conduct of the examination. He tried to examine all Ss in as nearly identical a manner as possible (see Appendix A).

For 16 hours before each session, Ss abstained from fluids other than milk, fruit juice, and water, and so began each session alcohol-free and caffeine-free. For the two treatment sessions, Ss arrived in a fasting state, having neither eaten nor drunk anything but water for five hours prior to these sessions. Ss neither ate nor drank (except the experimental drinks) nor smoked during any of the sessions.

All sessions for a given S were scheduled for approximately the same time of day, to minimize the effects of cyclical variations in body temperature, ambient temperature, and so on.

Experimental Treatments: Ss were assigned to one or four experimental treatments—a small dose of alcohol, a large dose of alcohol, caffeine, or placebo—according to a randomized blocks design.

The examination of Ss was spread out over a 14 month period. Ss were admitted to the experiment in chronological sequence; successive groups of four Ss were each designated a "block" on admission to the experiment. The four Ss of each block were assigned to the four experimental treatments at random.[1]

Ss were officially admitted to the experiment just before the third session (i.e., the first treatment session), as the experimental treatments were about to be assigned. The one participant who was unable to complete the two pre-treatment sessions was not assigned to an experimental treatment; the pre-treatment data obtained from this S were not considered part of the experiment. Results obtained from Ss assigned to treatments but unable to complete both treatment sessions were included in the study;

---

[1] Each block of Ss represented a miniature version of the experiment as a whole. The present randomized blocks design helped increase the efficiency of experimentation by segregating from experimental error the effects of unwanted variations (e.g., seasonal changes in ambient temperature) that occurred as the experiment progressed. For a brief introduction to randomized blocks, see, for example, Finney (1955); randomized blocks are further considered by Cochran and Cox (1957), by Fisher (1947), and by Maxwell (1958).

Design of the Study

Ss who dropped out during the treatment sessions were not replaced.

All Ss received pungent, grape-colored drinks. Drinks assigned to any given S contained one and only one of the following four doses:

1. Ethyl alcohol, 15.7 ml. absolute per square meter of body surface area;
2. Ethyl alcohol, 31.4 ml. absolute per square meter of body surface area;

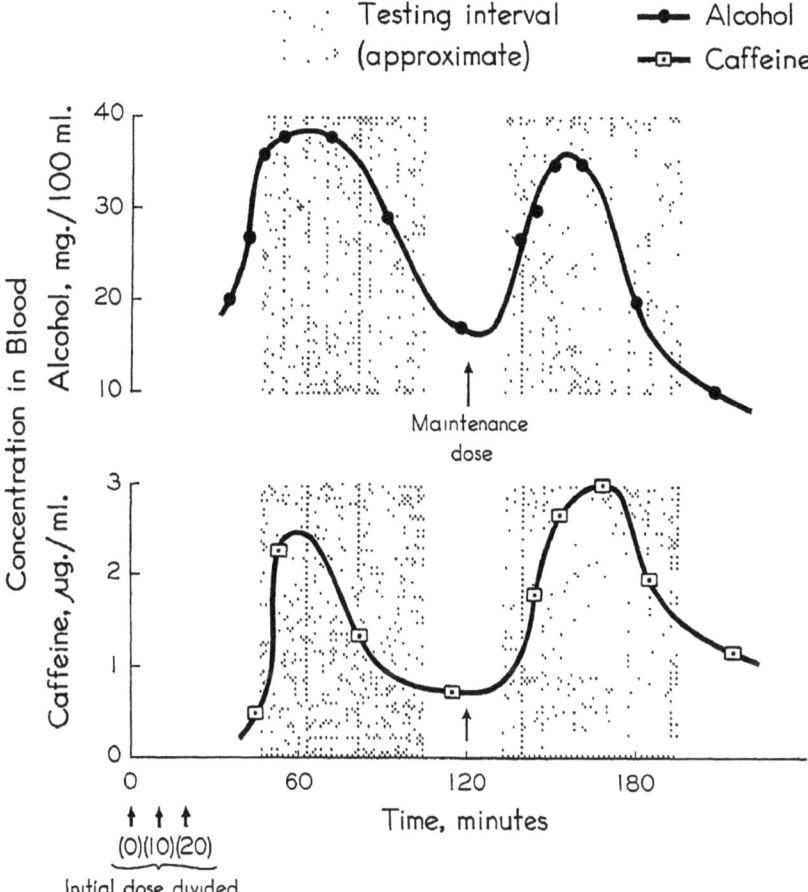

Fig. 1. Blood levels as a function of time, for the small alcohol and caffeine doses. Sample curves obtained in special sessions following completion of the experiment proper.

3. Caffeine alkaloid, 100 mg. per square meter of body surface area;

4. Neither alcohol nor caffeine (i.e., placebo).

Drinks were administered during the third and fourth sessions only, toward the beginning and toward the middle of the session.

The initial dose at each treatment session was divided into three portions administered at ten-minute intervals (see Table 1 and Fig. 1). Ss were encouraged to drink each portion in less than five minutes. No tests were administered until three-quarters of an hour after Ss began the initial drink. These procedures were followed to avoid too sharp a rise and fall in the drug concentration of the blood, and to avoid testing during the period of rapidly accelerating blood alcohol concentration (cf. Mellanby, 1919).

A maintenance or booster dose was administered about two hours after the initial dose, in order to replenish the quantities of drug which had been metabolized. The maintenance dose was given as a single drink. A half-hour test-free interval was allowed after ingestion of the maintenance dose was begun.

Alcohol doses were given in a solution composed, by volume, of 20 per cent absolute alcohol, 20 per cent water, and 60 per cent grape juice, to which was added 0.03 (0.04, for the first few blocks of Ss examined) ml. of peppermint oil per square meter of body surface area. Placebo Ss received the same initial and maintenance drinks as Ss receiving the smaller dose of alcohol, except that the alcohol was replaced by an equal volume of water. Except for the addition of caffeine, the initial and maintenance drinks received by caffeine Ss were identical to those received by placebo Ss. (Drinks were made up according to this formula to foster double-blind conditions; this is explained in Appendix A.)

Ss receiving the smaller alcohol dose were given a maintenance drink containing $(5.20A - 2.02)T$ ml. of absolute alcohol, while Ss receiving the larger alcohol dose were given a maintenance drink containing $(3.84A + 1.78)T$ ml. of absolute alcohol. Caffeine Ss received a maintenance drink containing $25AT$ mg. of caffeine. (A represents S's body surface area, in square meters. T represents the number of hours, less a quarter of an hour, elapsed between the time S began ingesting the initial drink and the time S began the maintenance drink.)

Blood Samples: Ten blood samples were obtained by finger puncture with a disposable blood lancet during each treatment session. Blood samples were obtained when testing was begun, when testing was interrupted for the maintenance dose, when testing was resumed, and when testing was completed (see Table 1). Punctures were confined to the *non-writing* hand, except at session's end.

Blood samples were analyzed for their alcohol content by a micromethod adapted from Kent-Jones and Taylor (1954), using the microdiffusion method of Conway (1939). Concentrations of alcohol ranging from 0 to 80 mg. per 100 ml. of blood were measured with a standard deviation of approximately 1.5 mg. per 100 ml. of blood. A full complement of blood samples was obtained from all Ss irrespective of experimental treatment, in order that all Ss be treated alike. Only a fraction of the blood samples obtained from caffeine and placebo Ss were analyzed for alcohol content (for check purposes).

Development of a precise micromethod for analyzing the caffeine content of blood could not be completed in time for use in the study.

Blood samples were analyzed by the examiner's assistant, who also mixed and administered the experimental drinks, punctured Ss' fingers to obtain blood samples, and rated Ss' Reaction to Puncture (see Chapter 3). Although blood alcohol determinations and Reaction to Puncture ratings were not safeguarded by the double-blind procedures governing collection of the remaining data (see Appendix A), a determined effort was made to follow identical procedures for all Ss.

3 | PSYCHOLOGICAL PROCEDURES

PSYCHOLOGICAL procedures were chosen to measure a variety of higher mental functions, as well as other selected psychological variables. Included were tests of intellectual capacity expected to provide a challenge for the intellectually superior Ss serving in the study. Most tests were adapted from existing experimental or commercial tests; they were administered and scored in the standard manner, except where deviations from standard procedure are noted. Responses to tests were *written*, however, even where this was not standard procedure, *unless otherwise indicated*.

Identical or alternate forms of the tests presented during the treatment sessions were, in most instances, also presented during the pre-treatment sessions, in order to provide baselines against which the treatment measures could be compared. Alternate forms were commonly designated "A" and "B," or "I" and "II." Forms "A" and "I" were presented during pre-treatment sessions, while forms "B" and "II" were presented during treatment sessions, unless otherwise indicated. Some relatively unreliable tests were administered, during treatment sessions only, when it seemed desirable to *explore* the effects of the drugs on a function for which precise measures were lacking.

Brief descriptions of the tests, grouped according to test battery, are presented below. The tests, listed in the order in which they were administered during pre-treatment sessions, are numbered consecutively to facilitate later reference. A dagger † designates tasks presented in identical form during pre-treatment and treatment sessions. An asterisk * designates speeded tests, in which Ss were encouraged to complete the largest possible number of items during a limited time interval.

Sessions 1 (Pre-treatment) and 3 (Treatment):

*† *(1) Writing Speed (This procedure is not part of a battery. It was administered during sessions 1 and 3 immediately before the tests of batteries A, B, C, D, and E.):* Write the phrase "How now brown cow" as rapidly as possible. Keep repeating the phrase for one minute. (Cf. Downey, 1923.)

Scores: Number of letters written. Also determined was † Writing Expansiveness, i.e., the average length (in millimeters) of the phrases completed without interruption. Scores were obtained for the entire one-minute period and for the first and second thirty-second periods.

BATTERY A

† *(2) Digits Forward:* Recall lists of digits immediately after presentation of each list. A set of three lists was presented for each of seven difficulty levels, making a total of twenty-one different lists ranging in length from four to ten digits. All three lists of a given length were presented before proceeding to the next higher level of difficulty. The first of each of the seven sets of lists together constituted trial 1. Similarly, the lists presented second and third in each of the seven sets together constituted trials 2 and 3, respectively.

Lists were read aloud, with digits spaced at one-second intervals. Lists were obtained, with minor modifications, from the *Memoirs of the National Academy of Sciences* (1921); the first two trials are identical with those of the Wechsler Memory Scale, Form II (Stone, Girdner, and Albrecht, 1946).

Scores: A list reproduced without error was scored 1. No weight was assigned to other lists. Separate scores were obtained for each trial, and for all three trials combined.

† *(3) Digits Backward:* Recall lists of digits in reverse order to that in which they are presented. Except for the recording of digits in reverse order, and except for the use of lists (from three

to nine digits in length) differing from those of Digits Forward, the procedure was identical with that followed for Digits Forward.

(4) Abstract Reasoning: Discover the principles by which sequences of diagrams are generated. Demonstrate your understanding of the principles governing the changing diagrams by indicating the diagram which should follow next in each unfinished sequence. All odd items, plus items 46, 48, and 50, of the D.A.T. Abstract Reasoning Test (Bennett, Seashore, and Wesman, 1952) were presented. Time limit: 14 minutes.

Scores: The Abstract Reasoning protocols were scored for † Writing Intensity, as well as in the standard manner (i.e., number of correct answers minus one-fourth the number of incorrect answers). The average depth of impression produced by Ss' penciled responses for Abstract Reasoning was rated independently by two judges, along a scale ranging from zero (for protocols with the lightest pencil impression) to nine. The ratings of the two judges, which correlated +0.76, were summed to provide the Writing Intensity score (cf. Free Association).

BATTERY B

(5) Sentence Memory: Recall sentences immediately after each is read aloud. Two sentences at each of six levels of difficulty were presented; alternate test forms were administered in sessions 1 and 3. Sentences of from 12 to 28 syllables, included in the lists of sentences prepared by Wells (1927), were supplemented by specially prepared thirty-two-syllable sentences.

Score: One credit for sentences reproduced without error; half-credit for sentences with one error. *All* successes (including those above the level at which two sentences were failed) were credited.

* *(6) Match Problems:* Groups of "matches" are laid out to form patterns of squares and triangles; remove given numbers of "matches," leaving prescribed numbers of squares or triangles intact. Items 10 to 18 of Match Problems, Form CX03B (Berger, Guilford, and Christensen, 1957), were presented in session 1; items 1 to 9 were presented in session 3.

Score: Number of correct answers.

BATTERY C

* *(7) Consequences:* List as many consequences as you can of certain unusual changes that suddenly alter the normal state of affairs—changes capable of seriously disrupting the living conditions of many people. Presented in session 1 were items "coal and oil" and "books" of the Consequences Test, Form CF06A (Berger *et al.*, 1957; Guilford, Frick, Christensen, and Merrifield, 1957; Guilford, Wilson, and Christensen, 1952; Guilford, Wilson, Christensen, and Lewis, 1951; Kettner, Guilford, and Christensen, 1959). Presented in session 3 were items "mind" and "blind." Two minutes were allowed for response to each problem, after the examiner read the problem aloud twice.

Scores: Four scores were obtained in addition to the standard Low-quality and Remoteness scores. Responses were scored for Favorable Outcome or Unfavorable Outcome in situations marked by sudden deprivation. "Mind," which differed from the other three items in not emphasizing deprivation, was not scored for outcome. Scores for "blind" were therefore doubled, to make them comparable to the combined outcome scores for "coal and oil" and "books."

Responses were also scored for Active Efforts at Mastery, or Passivity, in response to sudden deprivation. Scores for "blind" were doubled, since "mind" was not scored.

Subtraction of Low-quality from Passivity yielded a Passivity, Corrected score.

* *(8) Addition:* Add sets of four two-digit numbers as quickly as possible. Sixty-three-item alternate forms were prepared, with the aid of a table of random numbers, for sessions 1 and 3. Time limit: 5 minutes.

Scores: Number of correct answers; number of incorrect answers.

* *(9) Arithmetic Errors:* Discover the *one* change in *sign* needed to eliminate the error in expressions such as: $3 - 1 = 1 \times 1$. Sixteen-item alternate forms of this modification of the Sign Changes tests (Kettner *et al.*, 1959) were prepared for sessions 1 and 3 (see Appendix B). Time limit: 90 seconds.

Score: Number of correct answers.

* *(10) Controlled Associations:* List three synonyms for each of ten common words. Presented in session 1 was Part II of Controlled Associations II, Form CAF02B (Berger *et al.*, 1957; Kettner *et al.*, 1959). Part I was presented in session 3.

Scores: Total number of associations; Synonyms (number of acceptable synonyms among the associations offered).

BATTERY D

* *(11) Plot Titles:* Devise as many clever titles as you can for brief, fanciful tales. Presented in session 1 were plots "wise man" and "quack" of Plot Titles, Form CF02C (Guilford and Christensen, 1956; Guilford *et al.*, 1957; Kettner *et al.*, 1959). Presented in session 3 were plots "missionary" and "talking wife."

Scores: The standard Low-quality and Cleverness scores were obtained.

(12) Deduction: Given sets of statements presented in syllogistic form, indicate whether the conclusions necessarily follow from the premises. Subtest 3 (Deduction) of the Watson-Glaser Critical Thinking Appraisal (Watson and Glaser, 1952) was presented. Time limit: 9 minutes.

Score: Number of correct answers minus number of incorrect answers. (This differs from the score [number of correct answers] recommended by Watson and Glaser.)

BATTERY E

(13) Associate Learning: After hearing a list of word-pairs, recall the second word of a given pair when the first word is presented. The Associate Learning subtest of the Wechsler Memory Scale (Stone *et al.*, 1946; Wechsler, 1945) was modified by announcing the correct word-pair after each response and by omitting the third trial.

Scores: Number of Easy Associations and number of Hard Associations recalled correctly on the two trials.

* † *(14) Free Association:* Record as many different words as you can in two minutes. Record whatever single words come to mind (cf. Terman and Merrill, 1937).

Scores: Number of words for the entire two-minute period and for the first and second one-minute periods.

A † Writing Intensity score was also obtained (cf. Abstract Reasoning) by summing the ratings of two judges. The judges' ratings, which ranged from zero to seven, correlated +0.86.

* *(15) Word Fluency:* In three minutes, record as many different single words beginning with the letter "m" (in session 1; the letter "b" was used in session 3) as you can (cf. Thurstone, 1938).

Scores: Number of words beginning with the specified letter. Scores were obtained for the entire three-minute period and for the first and second ninety-second periods.

* † *Writing Speed was repeated (and* † *Writing Expansiveness was again determined) immediately after completion of the above tests.*

Sessions 2 (Pre-treatment) and 4 (Treatment):

BATTERY F *(tests of Battery F were administered individually in both pre-treatment and treatment sessions)*

* *(16) Street Gestalt Completion:* Identify and *call out* the names of objects or parts of objects, drawings of which are partly obliterated. Items from the four-page Street Gestalt Completion Test booklet (revised 1942 [unpublished]; cf. Thurstone, 1944), adapted by Thurstone from Street (1931), were employed. Alternate items from this booklet were segregated to provide two twelve-item alternate forms of the test, for use in sessions 2 and 4. Time limit: 15 seconds.

Score: Number of correct answers.

* *(17) Digit Symbol:* Substitute nonsense symbols for digits, using an arbitrary code that associates each of nine digits with a different symbol. Two consecutive trials of the Digit Symbol subtest of the Wechsler-Bellevue Intelligence Scale, Form II (Wechsler, 1946), were given in session 2. Four consecutive trials of the Digit Symbol subtest of the Wechsler Adult Intelligence Scale (Wechsler, 1955) were given in session 4. Time limit: One minute for each trial.

Scores: Number of items completed for each trial and for each thirty-second period. Errors, which were uncommon, were *not* taken into account in the scoring.

Before the third and fourth trials of session 4, Ss were asked to predict the number of items they expected to complete on the coming trial. Score *(for session 4 [treatment] only):* Attainment Discrepancy (Lewin, Dembo, Festinger, and Sears, 1944), i.e., the difference between the number of items actually completed and the number of items predicted, for the third and fourth trials combined.

Each S was asked to predict the greatest number of completed items attainable by himself and the greatest number attainable by the average adult. Score *(for session 4 [treatment] only):* Predicted Ceiling, $\frac{\text{Self}}{\text{Average}}$.

After the fourth trial, Ss were asked to estimate the number of items they had completed on each trial of session 4. Score *(for session 4 [treatment] only):* Estimation of Digit Symbol, Corrected, or the average difference (for the four trials) between the number of items estimated and the number of items actually completed.

† *(18) Stroop Color-Word:* This test is based on the work of Stroop (1935). Stimulus materials were presented on three cards prepared by the Psychometric Laboratory, University of North Carolina. The format of the three cards was similar. Each card was printed on a white background, and each contained 100 items —10 items per line.

Printed in black on Card *A* were the *names* of four colors: yellow, red, blue, and green. Printed on Card *B* were circular areas colored yellow, red, blue, or green. The colors on Card *B differed in sequence* from the names of colors on Card *A*.

Printed on Card *C, in color* rather than in black, were the four color names of Card *A*. Since the names on Card *C* were presented in the same sequence as the names on Card *A*, while the colors on Card *C* were presented in the same sequence as the colors on Card *B*, the color names on Card *C* were inconsistent with the colors in which they were printed. In fact, no name on Card *C* corresponded with the color in which it was printed.

Read aloud the color names on Card *A*, serially, as quickly as possible. Identify the colors on Card *B*, calling out the colors serially, as quickly as possible. Identify the *colors* on Card *C—disregard the color names* on this card—calling out the colors serially, as quickly as possible.

Scores: Time in seconds per 100 items, for each card. Scores were designated Stroop Reading Time, for Card *A;* Stroop Color-naming Time, for Card *B;* Stroop Conflict Time, for Card *C*. The ratio $\frac{\text{Stroop Color-naming Time}}{\text{Stroop Conflict Time}} \times 100$ was designated Stroop Ratio × 100.

* *(19) Continuous Subtraction:* Count backwards, aloud. (Cf. Sherman, 1924.) In session 2, count backwards from 73, by threes. In session 4, count backwards from 99, by fours. Time limit: 20 seconds.

Score: Number of correct subtractions.

BATTERY G[1]

(20) Story Memory: Immediate Recall: Recall stories immediately after hearing them read aloud. The Logical Memory Passages of the Wechsler Memory Scale (Stone *et al.*, 1946; Wechsler, 1945) were presented.

Score: Number of ideas reproduced immediately. Scores for the two stories of session 2 were *averaged;* similarly for session 4.

* *(21) Clerical Speed and Accuracy:* Locate given combinations of letters and/or numerals from among groups of similar combinations of letters and/or numerals. Part I as well as Part II of the D.A.T. Clerical Speed and Accuracy Test (Bennett *et al.*, 1952) was presented and scored.

Scores: Number of correct answers for each part.

* *(22) Gottschaldt Figures:* Indicate whether given designs are included (hidden) in more complex drawings. Items from the four-page Gottschaldt Figures Test booklet, arranged by Thurstone (unpublished; cf. Thurstone [1944]), were assigned to one of two thirty-six-item alternate forms of the test, for use in sessions 2 and 4. Time limit: 2 minutes.

Score: Number of correct answers minus number of incorrect answers.

---

[1] On completion of Battery F, in session 2 only, Ss' heights and weights were measured; body surface area (in square meters) was then determined (DuBois, 1936). See Table 1.

The Vocabulary section of the Cooperative Reading Comprehension Test C2 (Higher Level), Form Y (Bradford, Davis, Davis, Derrick, Neville, Spaulding, and Willis) was then administered. There were 60 five-choice Vocabulary items. Score: Number of correct answers minus one-fourth the number of incorrect answers. Raw scores were not converted to percentiles.

(23) Story Memory: Delayed Recall: Recall the stories heard earlier in session 4. (This task, of which Ss received no prior notice, was administered in *session 4 only*. The interval between Immediate Recall [20] and Delayed Recall, which varied slightly from S to S, was approximately 20 minutes.)

Score: The average number of ideas reproduced after delay, minus the average number of ideas reproduced immediately, for the two stories heard earlier in session 4.

BATTERY H

(24) Contingencies: Indicate possible conditions or circumstances that may arise requiring the use of given objects in specific situations. Presented in session 2 were situations 1 to 5 of Contingencies, Form PX03A (Berger *et al.*, 1957). Presented in session 4 were situations 6 to 10. (Contingencies' status as a power or speeded test is a bit uncertain, at least in the present study. Many Ss failed to complete the test, while many others finished well within the 6 minute time limit.)

Score: Number of acceptable answers.

* *(25) Mutilated Words:* Identify printed words, in which a part of each letter has been obliterated. Items from the four-page Mutilated Words Test booklet, Form 1950B, arranged by Thurstone (Pemberton, 1951) were employed. Alternate items from this booklet were segregated to provide two twenty-five-item alternate forms of the test. (Odd items, except the last, were presented in session 2; even items were presented in session 4.) Time limit: 2 minutes.

Score: Number of correct answers.

(26) Language Usage: Given sentences containing errors in grammar, punctuation, or spelling, indicate which components of the sentences are in error. The last 25 sentences of Sentences, Part II, of the D.A.T. Language Usage Test (Bennett *et al.*, 1952), were presented. Time limit: 13 minutes.

Score: Number of correct answers minus number of incorrect answers.

BATTERY I *(Administered during session 4 [treatment] only)*

(27) Picture Completion: Identify and call out the important feature omitted in a drawing of an object or scene. The Picture

Completion subtest of the Wechsler Adult Intelligence Scale (Wechsler, 1955) was presented. Time limit per item: None.

Score: Raw scores (number of correct answers) were used without conversion to weighted scores.

(28) Sentence Completion: Finish incomplete sentences as rapidly as possible, using a single word or phrase. Forty-seven incomplete sentences were adopted, with minor modifications, from published (Stein, 1947) and experimental forms of Stein's Sentence Completion Test. Time limit: 12 minutes.

Scores: Word Count is the total number of words written in response to the first 17 incomplete sentences.

Eleven items (3, 4, 9, 15, 18, 21, 25, 32, 38, 42, and 47; see Appendix C) were grouped into a Reaction to Stress "cluster" (see Nash, 1958). "Constructive" responses were scored +1; "unconstructive" responses were scored −1; intermediate responses were scored zero. The scores for the "stress" items, when totaled, yielded a Constructive Reaction to Stress score for the cluster as a whole. Responses for the "stress" cluster were also scored, in similar fashion, according to the alternative criterion: Mobilization of Aggressive Energies for Self-Forwarding Action (in Reaction to Stress).

The Reaction to Separation cluster, consisting of items 12, 23, 34, and 40 (see Appendix C), was scored according to the criterion: Constructive Reaction to Separation. Five scoring categories (with weights ranging from +2 to −2) were applied to the "separation" items, in contrast with the three categories used to score items of other clusters.

The Expression of Feelings cluster, consisting of items 6, 27, and 36 (see Appendix C), was scored according to the criterion: Favorable to Open Expression of Feelings.

Responses to the remaining Sentence Completion items were unscored, except for items included in Word Count.

S's response to any given item was scored without reference to other responses produced by the same S.

(29) Evaluation of Arguments: Distinguish strong arguments from weak or irrelevant arguments. Subtest 5 (Evaluation of Arguments) of the Watson-Glaser Critical Thinking Appraisal (Watson and Glaser, 1952) was administered; Forms AM and BM were combined into a single form. Time limit: 10 minutes.

Score: Number of correct answers minus number of incorrect answers. (This differs from the score [number of correct answers] recommended by Watson and Glaser.)

BATTERY J *(Administered during session 4 [treatment] only)*

(30) Practical Judgment: Selected items from Cardall's Test of Practical Judgment (Cardall, 1942) were administered. Time limit: 12 minutes.

Scores: The scale Understanding How to Handle an Emergency consisted of items 23, 36, 37, 41, and 44. The scale Understanding the Reasons for Accepted Social Practices consisted of items 25, 27, 33, 35, 38, 43, 46, plus Sample items A and B. The scoring weights developed by Cardall were applied to the individual items; arbitrary scoring weights were assigned to Sample items A and B.

(31) Slow Writing: Write the word "on" as slowly as possible. (Cf. Downey, 1923.)

Score: Duration of writing, in seconds.

(32) Rosenzweig Picture-Frustration: Two persons are depicted in mildly frustrating situations of common occurrence. Indicate the reply of one person to remarks made by the other. The Rosenzweig Picture-Frustration Study, Revised Form for Adults (Rosenzweig, Fleming, and Clarke, 1947), was administered. Time limit: 12 minutes.

Scores: Extropunitiveness versus Intropunitiveness is the *difference* between the total extropunitive score and the total intropunitive score. Expression of Aggression is the *sum* of the total extropunitive score and the total intropunitive score. Striving to Overcome Frustration is the sum of Rosenzweig's scoring factors "e" and "i." Scores are presented in terms of frequency of occurrence (frequencies were *not* converted to percentages).

Several procedures were administered during session 4 (treatment), on completion of Batteries F, G, H, I, and J, in the following order:

(33) Incidental Learning: As each of 14 two-digit numbers is announced, locate and circle the number on a sheet containing numbers 11 through 99. This ostensible coding task was repeated

four times, with minor variations. The subjects, who were unaware that their retention of the 14 numbers would be tested, were then asked to recognize these numbers. The test procedure was identical to that followed by Neimark and Saltzman (1953) for their Group II, except that: (a) auditory rather than visual presentation was employed—the reading of each of the 14 digit-pairs lasted for one second, and there was a two-second interval of silence between each of the two-digit numbers; and (b) five minutes were allowed for recognition.

Score: Number of digit-pairs correctly recognized, minus one-fourth the number of digit-pairs correctly or incorrectly recognized.

(34) Estimation of Treatment Effects: For each of five tests administered both in sessions 2 and 4, compare your performance under pre-treatment and treatment conditions. For each test, check one of the following categories (scoring weights are indicated in parentheses): (+2) much better in session 4; (+1) somewhat better in session 4; (0) no difference between sessions 2 and 4; (−1) somewhat better in session 2; (−2) much better in session 2.

Scores: Scoring weights for the five tests were totaled to yield an over-all estimate. After having their scales adjusted, the scores actually obtained for the five tests were subtracted from the over-all estimate to yield a corrected estimate of treatment effects.

(35) Whiskey Estimate: Estimate how much whiskey you would have to drink at once to feel like you did during session 4.

Score: Pints or fraction of a pint, expressed in per cent.

Whiskey Estimate, Prostration: Estimate how much whiskey you would have to drink at once to get into the same condition as people you see lying in the street after drinking too much alcohol.

Score: Pints or fraction of a pint, expressed in per cent. The ratio $\frac{\text{Whiskey Estimate}}{\text{Whiskey Estimate, Prostration}} \times 100$ is designated Whiskey Estimate, Corrected.

(36) Estimation of Slow Writing: Estimate how long it would take you to write "Northwestern University," as slowly as possible.

Scores: Number of seconds. Estimates corrected for actual Slow Writing values were also computed.

(37) Time Estimation: (I) Rank four tests, administered both

in sessions 1 and 3, according to the amount of time allowed for working on these tests. (II) Repeat, for four tests administered both in sessions 2 and 4. The time limits actually allowed for the eight tests being judged ranged from 20 seconds to 13 minutes.

Scores: Accuracy of time estimation was indicated, both for Task I and for Task II, by coefficients of rank correlation between rankings assigned by Ss and rankings of the time limits actually allowed for the tests.

† **Self-ratings:** Ss indicated their feelings and states of mind by rating themselves on ten variables. For each of the variables: Relaxed; Irritable; Satisfied with Myself; Physically Uncomfortable; Tired and Weary; Capable of Paying Close Attention; Impulsive, Lacking in Restraint; Keen and Alive; Interested in Taking the Tests; Would Trust My Judgment, S was asked to report "how you feel, right now." For each of the ten self-ratings, S was presented with seven response categories: not at all, very slightly, slightly, moderately, quite a bit, very much, and extremely. These seven categories were assumed to form an interval scale (see Jones and Thurstone [1955]) and were assigned scoring weights of 0, 1, 2, 3, 4, 5, and 6, respectively.

The foregoing ten variables were rated, consecutively, on thirteen specified occasions during the four sessions of the experiment. These ratings were made either toward the beginning, toward the end, or midway through the various sessions (see Table 1). Only those self-ratings obtained immediately before and immediately after administration of the psychological tests, during each of the four sessions, have been analyzed; self-ratings for the remaining five occasions are not considered. Ratings for each of the four post-drug occasions were corrected, by subtraction, for the corresponding pre-drug ratings.

Ss were also asked to indicate the degree to which they disliked the idea of having their fingers punctured for a blood test, using the same seven response categories referred to above. Ratings for Dislike Puncture were obtained on four occasions only (see Table 1). Those obtained at the end of the two treatment sessions have been corrected, by subtraction, for the two pre-treatment ratings.

Reaction to Puncture: Finger punctures made while obtaining blood samples produced mild but observable responses in a number of Ss. Two aspects of S's reaction were rated and analyzed: Tightening of the extended finger, while it was being held, preparatory to or during puncture; and Withdrawal of the hand or finger, on puncture.

Scores: Tightening and Withdrawal were rated, for each puncture, using five scoring categories: not noticeable, barely noticeable, mild, fair-sized, and strong. Scoring weights of 0, 1, 2, 3, and 4, respectively, were arbitrarily assigned to these categories.

Tightening ratings for both treatment sessions were summed to give a single Tightening score, since there were no marked changes in rating within or between sessions. A similar procedure was followed for Withdrawal. Both totals were prorated for missing values.

PART II
Results and Discussion

4

BACKGROUND INFORMATION

THE PRESENT chapter provides background information for the remaining chapters of Part II. The subject sample is described, and blood levels associated with the active treatments are indicated. The principal results of the experiment are discussed in Chapters 5, 6, and 7, according to psychological function; the present chapter indicates how the quantitative results were analyzed, and how the summary findings are tabulated. Results relating to the double-blind procedure are discussed in Appendix A. The extent to which psychological functioning varied with blood levels and with customary consumption of drugs is discussed in Chapter 8.

Subjects: Most Ss were male college students in their early twenties. The distribution of Ss' ages was skewed, with mean age 24.0 years exceeding median age 22.5 years; see Table 2. Ss' mean Vocabulary score exceeded that attained by the typical entering college freshman.[1] Body surface area exceeded that of the average adult, as would be expected from a predominantly male sample.

Alcoholic and caffeinated beverages were customarily consumed in moderation by most Ss. About one fifth of the Ss abstained from alcohol entirely, or had perhaps one or two glasses of wine a year. One atypical S consumed two dozen alcoholic drinks per

---

[1] See *Cooperative English test, Single booklet edition: Percentile ranks for high school and college students.*

TABLE 2
Characteristics of the Subject Sample

Measure	Mean for All 56 Ss	Placebo Mean	\multicolumn{3}{c}{Difference Between Means of Active Treatment and Placebo}
			CAFFEINE	ALCOHOL (15.7 ML.-/M.^{2})	ALCOHOL (31.4 ML.-/M.^{2})
Age (years)	24.00	23.71	0.54	0.61	0.00
Vocabulary (raw scores)	42.52	38.07	7.57	6.36	3.86
Body surface area (square meters)	1.84	1.77	0.15	0.04	0.08
Alcoholic drinks customarily consumed (per week)	2.47	2.59	−0.77	1.16	−0.87
Coffee-cup equivalents of caffeine customarily consumed (per day)	2.72	2.48	0.54	−0.13	0.57

week, but the median number of alcoholic drinks consumed was one per week.[2]

All Ss drank one or another caffeinated beverage, at least occasionally. One S consumed ten cups of coffee a day. Customary caffeine consumption was, however, less skewed in its distribution than was customary alcohol consumption. The median amount of caffeine consumed per day was equivalent to that in one and a half cups of coffee.[3]

Ss receiving the active treatments failed to differ significantly from Ss receiving placebo with respect to any of the background characteristics presented in Table 2. It is worth noting, however, that the pre-treatment intellectual level of both caffeine Ss and smaller alcohol dose Ss surpassed that of placebo Ss. The mean Vocabulary score for caffeine Ss in fact exceeded that for placebo Ss by 7.57 units, a value close to the threshold of significance: 8.33 units was the 2-sided, 0.95 Dunnett allowance (Dunnett, 1955;

---

[2] A beverage containing a jigger of whiskey was taken as the unit of measurement in estimating customary alcohol consumption. Drinks having a lower alcohol content, such a glass of wine, were assigned fractional credit.

[3] A cup of tea was considered equivalent to a cup of coffee, which was adopted as the unit of measurement in estimating customary caffeine consumption. Milder caffeine drinks received fractional credit (four regular-sized cola beverages were equated with one cup of coffee, for example).

see the section: *Analysis of data,* below). Caution is therefore advised in interpreting any findings suggesting that caffeine or the smaller alcohol dose enhanced intellectual functioning, if these findings were not adjusted for pre-treatment differences. (See the section: *Associative productivity,* Chapter 6; also see Chapter 7.)

Of the 56 Ss assigned to treatments, six Ss who received the heavy alcohol dose failed to complete the experiment. Data obtained before illness struck these six Ss—who experienced nausea or emesis—were included in the analysis of results.

The tests omitted by the Ss who failed to complete the experiment varied from S to S. These Ss became ill at different times during sessions 3 and 4, but all of them failed to take the non-battery tests administered toward the end of session 4 (see Table 1 and Chapter 3). Since session 4's end-of-session measures were mostly indices of critical judgment, the illness of the six Ss reduced the effectiveness with which heavy alcohol dose effects on critical judgment were measured (see the section: *Critical judgment,* Chapter 6). The measurement of heavy alcohol dose effects on other psychological functions was not markedly disrupted by the drop-outs, for the great majority of tests were administered in batteries, whose order of presentation was randomized. Heavy alcohol dose sample size ranged from a minimum of eight Ss for some measures to a maximum of fourteen Ss for others.

Blood Levels for Alcohol and Caffeine: The maintenance dose succeeded in keeping blood alcohol levels within a range of about 15 to 60 mg. per 100 ml. of blood for the small alcohol dose, and within a range of about 45 to 90 mg. per 100 ml. of blood for the heavy alcohol dose, *during the course of post-drug testing.* Noteworthy are the rather low blood alcohol levels at which tests were occasionally administered, following ingestion of the small alcohol dose.

Peak blood levels tended to be reached at about the time the third finger-tip blood sample was obtained (see Table 1), toward the beginning of the post-drug sessions, and at about the time the eighth finger-tip sample was obtained, following the maintenance dose. Blood levels tended to reach a low point, while testing was in progress, at about the time the fourth/fifth (pre-maintenance dose) and ninth/tenth (post-maintenance dose) finger-tip samples were obtained.

For both alcohol doses, and for both post-drug sessions, peak levels attained following the maintenance dose were approximately equivalent, on the average, to those attained toward the session's start. Fig. 1 presents a sample blood alcohol curve (see the section: *Experimental treatments*, Chapter 2), whose values were obtained after completion of the experiment proper. These sample values are spaced more evenly than the blood alcohol values obtained during sessions 3 and 4. (Sample values are presented in Fig. 1, rather than actual values, because they permit the drawing of a smoother curve. For blood alcohol levels actually observed during the experiment, see Figs. 4 and 5 in Chapter 8.[4])

Blood caffeine levels were not determined during the experiment proper, but sample blood values were obtained following the completion of the experiment, using a macromethod developed by Axelrod and Reichenthal (1953). Blood caffeine levels appeared to range between 0.5 and 5.0 mg. per liter of blood. A sample blood caffeine curve is also presented in Fig. 1.

The maintenance schedules for both alcohol and caffeine apparently restored blood levels, at least approximately, to their pre-maintenance-dose values. The maintenance schedules also helped limit the variation of blood levels during a given session.

Analysis of Data and Tabular Presentation of Results: Post-drug values (corrected by subtraction for the corresponding pre-drug values, whenever the latter were available) were subjected to a two-criterion analysis of variance. (For the analysis of a two-way experiment, see Snedecor [1956].) Estimates for missing data were calculated and included in the analysis (also see Snedecor). Block effects were rarely significant; they are indicated only when a full analysis of variance is presented (see Table 5 in Chapter 5, and Tables 8, 10, 12, and 14 in Chapter 6).

---
[4] Fig. 5 charts Writing Expansiveness values observed 48 minutes after the start of session 3 against blood alcohol values observed six minutes earlier; see Table 1. Fig. 4 is based on the writing trial conducted toward the end of session 3; Writing Speed values observed at this time are plotted against blood alcohol levels observed immediately afterwards.

TABLE 3
Composite Measures[a]

Composite Measures	Factors for Component Measures	Component Measures
Adaptive Flexibility	1 1 2	Arithmetic Errors (9)[b] Gottschaldt Figures (22) Match Problems (6)
Associative Productivity	2 2 1 5 5	log ([Consequences: Low-quality] + 10) (7) log ([Controlled Associations: Synonyms] + 10) (10) log ([Plot Titles: Low-quality] + 2) (11) log (Free Association + 50) (14) log (Word Fluency + 50) (15)
Consequences-Plot Titles	2 2 1 2	log ([Consequences: Low-quality] + 10) (7) log ([Consequences: Remoteness] + 10) (7) log ([Plot Titles: Low-quality] + 2) (11) log ([Plot Titles: Cleverness] + 10) (11)
Constructive Reaction to Stress	2 1 2	Rosenzweig Picture-Frustration: Striving to Overcome Frustration (32) Sentence Completion: Constructive Reaction to Stress (28) Sentence Completion: Reaction to Separation (28)
Immediate, Intentional Recall	3 3 3 4 2	Associate Learning: Easy (13) Associate Learning: Hard (13) Digits Forward (3 trials) (2) Sentence Memory (5) Story Memory: Immediate Recall (20)
Practical Judgment	2 4 1 1	Evaluation of Arguments (29) Picture Completion (27) Practical Judgment: Understanding How to Handle an Emergency (30) Practical Judgment: Understanding the Reasons for Accepted Social Practices (30)
Reaction to Puncture	1 1	log ([Reaction to Puncture: Withdrawal] + 2) log ([Reaction to Puncture: Tightening] + 2)
Visual Adaptive Flexibility	1 2	Gottschaldt Figures (22) Match Problems (6)
Visual Thinking	5 10 2 10	Gottschaldt Figures (22) Mutilated Words (25) Perceptual Speed Composite (21) Street Gestalt Completion (16)
Writing Intensity	1 2	Writing Intensity (Abstract Reasoning) (4) Writing Intensity (Free Association) (14)

[a] Tabulated here are those composite measures not fully described in the text. A composite measure was obtained by summing the products of its component measures and their scale factors.

[b] Numbers in parenthesis at the *end* of a line have no computational significance. These are the numbers assigned tests in Chapter 3 to facilitate reference to test descriptions.

Independent measures of a given function were combined, on occasion, to form a single composite measure. When measures displayed marked heterogeneity of variance, their scales were adjusted, before summation, to equalize their variances. The composition of measures having components with unequal scale factors is indicated in Table 3.

Composite measures were analyzed in essentially the same way that individual measures were. Estimates of missing data computed for components of a composite measure were included in the composite analysis. Degrees of freedom for the drugs × blocks interaction term were reduced by the number of Ss lacking data for *all* components of a given composite measure. Interactions involving blocks and sub-plot factors were rarely found to be significant (see the discussion of split-plot analyses by Anderson and Bancroft [1952] and by Brownlee [1960]); these interactions were therefore all treated as part of the residual (sub-plot) error term.

Presented in Tables 4, 7, 9, 11, 13, and 15 through 20 (see Chapters 5, 6, and 7) are comparisons between post-drug values (from which corresponding pre-drug values were subtracted, if available) for active treatments and placebo. Printed in roman type are the differences between the three active treatment means and the corresponding placebo means. Placebo means are listed separately, also in roman type, in order to indicate the baselines for the various comparisons. All mean values include the computed estimates for missing data, except for measures evaluated by non-parametric methods (see below). Composite findings are presented in terms of *average values per weighted component*.

Comparisons between active treatment means and corresponding placebo means were evaluated by reference to the two-sided, 0.95 Dunnett (1955) allowances (half-confidence intervals) presented in *italics* in Tables 4, 7, 9, 11, 13, and 15 through 20; each allowance is placed immediately below the corresponding comparison. Comparisons exceeding their allowances are significant[5]

---

[5] There is a one-to-one relationship between classical confidence intervals and tests of hypotheses. For example, if a (1-α) confidence interval excludes the value ξ of the parameter, then a test of the null hypothesis that the parameter has the value ξ will be rejected at the α level of significance. The Dunnett confidence intervals are designed for comparison of several (k) treatments with a

Background Information

at the 0.05 level (two-tailed test); significant comparisons are starred*.

Comparisons between means for active treatments and placebo, and Dunnett allowances corresponding to these comparisons, are presented for composite measures as well as for individual measures. For composite measures lacking significant drugs × components interactions, comparisons and allowances are the only findings presented, since they are the only findings of interest. For composite measures whose components interacted significantly with drugs, sub-plot findings are of interest, and full analyses of variance are presented (see Tables 5, 8, 10, 12, and 14).

Individual or composite measures displaying marked anormality or heterogeneity of variance were transformed to (common) logarithms (see, for example, Snedecor; see Bartlett [1947] for a more technical discussion). Data unsuitable for transformation were evaluated by means of the H-test (Kruskal and Wallis, 1952). Missing data were not estimated for measures evaluated by the H-test.

---

control, and have the property that on (1-α) of the occasions on which a set of such confidence intervals are constructed, all (k) of the intervals will be correct; but on α of the occasions, one or more of the intervals will be incorrect. These confidence intervals will be related to corresponding tests of hypotheses in a manner analogous to classical confidence intervals.

5 | SELF-RATING OF SUBJECTIVE STATES

FOR EACH of the ten variables rated periodically by Ss, beginning and final values for sessions 1 and 2 were subtracted from the corresponding post-drug values for sessions 3 and 4. The four difference scores for each variable were then totaled to yield a single composite self-rating. The results for these composite self-ratings (which include values for only eight of the thirteen occasions on which Ss rated themselves) are presented in Table 4. Also included in Table 4 are the findings for Dislike Puncture, a composite self-rating based on only two post-drug and two pre-drug values.

The self-ratings were hardly affected by caffeine. Temporal variations in caffeine effects were observed for one self-report measure, Tired and Weary; see the significant interaction of drugs and sessions, Table 5. Because of this interaction, Table 4 presents Tired and Weary findings for each treatment session. Table 4 indicates that caffeine reduced, or perhaps allayed, feelings of fatigue in session 4 (this finding is of borderline significance). That caffeine effects on feelings of fatigue were most evident during session 4 is not too surprising, considering that the increase in feelings of fatigue between sessions 2 (the shortest of the four sessions) and 4 (the longest session) exceeded that between sessions 1 and 3. Apparently, the more marked the fatigue, the greater its susceptibility to caffeine's influence. Feelings of fatigue reported by Barmack's (1940) Ss were allayed by a caffeine dose

TABLE 4
Treatment Effects on Self-ratings[a]

Measure	Placebo Mean	CAFFEINE	ALCOHOL (15.7 ML.-/M.²)	ALCOHOL (31.4 ML.-/M.²)
Relaxed	−0.11	0.41 / 0.75	0.27 / 0.75	0.34 / 0.75
Irritable	0.11	−0.16 / 0.56	0.36 / 0.56	−0.30 / 0.56
Satisfied with Myself	−0.13	0.23 / 0.71	0.07 / 0.71	0.29 / 0.71
Physically Uncomfortable	−0.04	0.02 / 0.66	0.23 / 0.66	0.14 / 0.66
Tired and Weary, session 3	0.14	0.04 / 1.39	0.46 / 1.39	−0.04 / 1.39
Tired and Weary, session 4	0.71	−1.21 / 1.26	−0.18 / 1.26	0.32 / 1.37
Capable of Paying Close Attention	−0.32	−0.02 / 0.76	−0.18 / 0.76	−0.75 / 0.76
Impulsive, Lacking in Restraint	0.09	−0.29 / 0.72	0.45 / 0.72	0.52 / 0.72
Keen and Alive	−0.50	0.36 / 0.86	0.21 / 0.86	−0.02 / 0.86
Interested in Taking the Tests	−0.36	0.14 / 0.67	−0.01 / 0.67	−0.04 / 0.67
Would Trust My Judgment	−0.36	0.07 / 0.80	−0.50 / 0.80	−0.90* / 0.80
Dislike Puncture	−0.07	1.00 / 1.35	1.00 / 1.35	0.71 / 1.42

[a] Four post-drug values were averaged for each self-rating measure, except for Tired and Weary, session 3 (*two* post-drug values averaged) and session 4 (*two* post-drug values averaged). *Two* post-drug values were *summed* for Dislike Puncture. Post-drug self-rating values were corrected by subtraction for corresponding pre-drug values.

[b] Two-sided, 0.95 Dunnett allowances are presented in *italics* directly beneath the corresponding differences between means. Comparisons significant at the 0.05 level are starred*.

TABLE 5

ANALYSIS OF VARIANCE OF DRUG EFFECTS
ON SELF-RATINGS FOR TIRED AND WEARY

Source of Variation	df	Mean Square	F	p
D (Drugs)	3	6.81	1.46	NS[a]
B (Blocks)	13	2.84	0.61	NS
D × B	39	4.67		
S (Sessions)	1	1.97	0.88	NS
BE (Beginning vs. end of session)	1	0.76	0.34	NS
S × BE	1	2.79	1.25	NS
D × S	3	7.06	3.17	0.05
D × BE	3	0.61	0.27	NS
D × S × BE	3	0.12	0.05	NS
Residual	145	2.23		

[a] Failed to reach significance at the 0.05 level.

only slightly smaller than that of the present study; the longer these Ss worked at a boring task, the more marked was caffeine's effect.

Barmack also observed that feelings of sleepiness and inattentiveness reported by his Ss were allayed by caffeine, though perhaps less markedly than feelings of fatigue. Barmack's Ss became sleepier and less attentive the longer they proceeded with their boring task; caffeine acted to forestall the sleepiness and the inattentiveness. Related self-ratings included in the present study—Keen and Alive, and Capable of Paying Close Attention—were, by contrast, hardly affected by caffeine.

Barmack further observed that feelings of boredom reported by his Ss were allayed by caffeine as they continued to work at their task. This also contrasts with present findings: self-ratings for Interested in Taking the Tests were essentially unaffected by caffeine.

How to account for the subjective caffeine effects obtained by Barmack but essentially undetected in the present experiment? Perhaps one or both sets of findings is in error?[1] Notwithstanding

---

[1] Neither Barmack's findings nor those of the present study are above question. Barmack's findings, for example, were evaluated by an insufficiently conservative

the possibility of such error, the author is inclined to believe that Barmack's Ss actually experienced different subjective caffeine effects than did Ss of the present study, the difference in subjective effects reflecting the different conditions under which Ss were examined. Barmack's Ss worked at a boring and fatiguing task; they had ample opportunity to experience the refreshing and restorative effect of caffeine. The present Ss remained fresher and more interested in what they were doing; there was less opportunity for caffeine's restorative effect to make itself felt.

It is here assumed that c.n.s. stimulants such as caffeine act primarily to *restore deteriorated feelings* rather than to *enhance normal feelings*. It is further suggested, as a corollary, that the less stimulating the environment, the more readily observable the subjective effects of c.n.s. stimulants such as caffeine.

Findings for methyl-caffeine (a caffeine derivative) obtained in Payne and Hauty's (1954) study of drug effects on pursuit performance have a possible bearing on the present discussion. Methyl-caffeine improved Ss' task disposition, as reflected by a combination of five questions pertaining to Ss' feelings of fatigue and boredom, to Ss' willingness to volunteer for another experiment, and to Ss' evaluation of the experiment's usefulness and of the medication's effect on pursuit performance. Enhancement of subjective state by the caffeine derivative was moderate compared to that induced by d-amphetamine and was due largely to Ss' recognition of methyl-caffeine's beneficial effect on pursuit performance.

Caffeine's subjective effects appear to be quite limited, except after prolonged fatigue or boredom of a degree not experienced in the present study.

---

test of significance; but this could hardly account for the *marked* differences between findings of the two studies. A further question about Barmack's findings is raised by the discrepancy between subjective caffeine effects obtained in two of Barmack's experiments.

The present dearth of subjective caffeine findings might, on the other hand, be attributable to imprecision of the seven-point unipolar self-rating scales (Barmack used nine-point bipolar scales). Subjective state was generally quite favorable at the start of the present experiment as well as after administration of placebo; might the present experimental conditions have limited the expression or impeded the detection of possible beneficial subjective caffeine effects?

Alcohol effects on the self-ratings were little more noticeable than those of caffeine. Alcohol's only clearly significant subjective effect was the heavily alcoholized Ss' expression of a reduced confidence in their judgment. Although the over-all intellectual capacity of Ss given the heavy alcohol dose had been considerably impaired (see below), these Ss were yet able to maintain sufficient self-critical capacity to be both aware and wary of their impaired condition. (See the section: *Critical judgment,* in Chapter 6.)

Ss also noted a reduced capacity to pay attention following ingestion of the heavy alcohol dose. This finding, reflecting the well-known hypnotic qualities of alcohol in large doses, was of borderline significance. Self-ratings for Keen and Alive were unaffected by alcohol, however.

A greater impulsiveness and a lessened restraint was reported for both alcohol doses. Although not significant, this finding suggests feelings of reduced self-control.

That Table 4 contains so few statistically significant findings is indeed surprising. It seems unlikely that the heavy alcohol dose, for example, had such a dearth of subjective effects. The Ss under heavy alcohol dosage certainly offered many spontaneous comments, usually in a negative vein, about their subjective state. Had the self-ratings been more precise research tools, significant findings might likely have resulted more frequently. Consider, for example, Dislike Puncture, which was rated less frequently than the remaining items of Table 4. Results for Dislike Puncture might well have been significant, had Ss had further opportunity to report their feelings about the blood-taking procedure. (See the section: *Reaction to bodily threat,* in Chapter 7.)

6

INTELLECTUAL FUNCTIONING

CAPACITY MEASURES, Over-all Findings: Presented in Tables 7, 9, 11, 13, and 15 through 20 are the principal findings for all measures other than the self-ratings. These measures, considered here according to the psychological functions to which they are most relevant, are in most instances indices of intellectual capacity. (When S is motivated to perform a given task to the very best of his ability, a measure of S's attainment of his desired objective is said to provide an index of S's capacity.) Also included in the tables mentioned above are expressive and projective measures (designated by superscripts [e] and [p], respectively), most of which are concentrated in Tables 7 and 20.

The question is here raised whether intellectual capacity was affected by the active treatments in an *over-all* way, regardless of effects on *specific* functions.

Inspection of Tables 7, 9, 11, 13, and 15 through 19 reveals that capacity findings were predominantly positive for caffeine and predominantly negative for the heavy alcohol dose. Capacity findings for the mild alcohol dose were, on the other hand, almost equally divided in sign. Since a positive sign indicates improved functioning for the great majority of capacity measures, and a negative sign, impaired functioning, it is plain that *over-all intellectual efficiency was enhanced by caffeine, impaired by the heavy alcohol dose, and affected little (if at all) by the mild alcohol dose.* If one ignores those few capacity measures for which adequacy of functioning is not revealed by sign (Whiskey Esti-

TABLE 6

SIGNIFICANCE AND DIRECTION OF DRUG FINDINGS FOR CAPACITY MEASURES[a]

	CAF-FEINE	Differences Between Active Treatment Means and Corresponding Placebo Means ALCOHOL (15.7 ML.-/M.²)	ALCOHOL (31.4 ML.-/M.²)
Types (and Number) of Measures			
Percentage of measures with statistically significant findings:			
34 independent measures *corrected* for baseline (i.e., pre-drug) values	3	0	12
6 independent measures *uncorrected* for baseline (i.e., pre-drug) values	0	0	0
40 independent measures, irrespective of correction for baseline values	3	0	10
7 composite measures	29	0	50
Percentage of measures indicating enhanced functioning:			
34[b] independent measures *corrected* for baseline (i.e., pre-drug) values	79	56	15
6 independent measures *uncorrected* for baseline (i.e., pre-drug) values	40	50	67
40 independent measures, irrespective of correction for baseline values	74	55	23
7 composite measures	86	43	0

[a] Summarized here are findings for all but those few capacity measures whose sign does not indicate adequacy of functioning. Where two or more measures lacked experimental independence, only one of these measures was included in the summary. Only the most inclusive of overlapping composite measures was included in the summary.

[b] Enhanced functioning is indicated by positive values for 29 of these measures, and by negative values for the remaining five measures.

mate [35],[1] for example, or Estimation of Digit Symbol, Corrected [17]), and if one distinguishes from the remaining capacity measures those few for which improvement is indicated by a *negative* rather than by a positive sign (Stroop Reading, Color-naming, and Conflict Times [18]; *log* Slow Writing [31]; Addition, Incorrect [8]), the effects of caffeine and of the heavy alcohol dose on over-all intellectual functioning stand out even more plainly.

The above impressions are confirmed by Table 6, which sum-

---

[1] The number in parentheses following a measure is that assigned to the corresponding test in Chapter 3.

marizes the capacity findings of Tables 7, 9, 11, 13, and 15 through 19. Table 6 is limited to capacity measures whose *sign* clearly indicates whether functioning has been enhanced or impaired by treatment. From Table 6, it is clear that most capacity measures were enhanced by caffeine. An even greater number of capacity measures was impaired by the heavy alcohol dose (subtracting from 100 the percentage of measures enhanced by a drug yields the percentage of impaired measures).

Approximately equal numbers of capacity measures were enhanced and impaired by the mild alcohol dose. To this finding, add the *complete* absence of significant findings for the mild alcohol dose, and it appears evident that the mild alcohol dose had no over-all effect on intellectual capacity.

It is noteworthy that the number of *independent* measures significantly affected by the heavy alcohol dose, and especially by caffeine, is hardly in excess of what would be expected by chance. That this dearth of significant findings was due to excessive experimental error is suggested by the fact that capacity measures *uncorrected* for baseline (i.e., pre-drug) values (and consequently providing *less precise* measurement) failed to reveal the over-all effects of caffeine and of the heavy alcohol dose so clearly evident from the capacity measures *corrected* for baseline values (see Table 6). That the actual effects of caffeine and of the heavy alcohol dose were *under*estimated (due perhaps to insufficient sample size or to insufficiently precise measurement) is further suggested by the fact that caffeine-induced enhancement and heavy alcohol dose-induced impairment of function were revealed more readily by composite than by individual capacity measures.

The percentage of significant findings reveals an even sharper contrast between composite and individual capacity measures. Providing greater precision of measurement than individual capacity measures, the composite measures most clearly reveal that intellectual functioning was improved by caffeine and disturbed by the heavy alcohol dose.

The foregoing analysis indicates that the findings for the heavy alcohol dose were generally similar in magnitude, though opposite in direction, to those for caffeine. But the actual potency of the heavy alcohol dose relative to caffeine is surely underestimated by these findings, since error estimates for the heavy alcohol dose increased as the number of heavily dosed Ss was reduced by illness.

There follows a more detailed examination of the capacity findings, according to psychological function.

Motor Control, Capacity Measures: Drug effects on Ss' capacity for deliberate and conscious motor control were investigated by measures reflecting the ease with which Ss were able to accelerate or retard their handwriting pace, at will.

TABLE 7
TREATMENT EFFECTS ON MOTOR CONTROL[a]

Measure	Placebo Mean	CAFFEINE	ALCOHOL (15.7 ML.-/M.²)	ALCOHOL (31.4 ML.-/M.²)
Writing Speed (1)[b]	−2.25	1.77 5.00	0.66 5.00	−3.39 5.12
Writing Expansiveness, beginning of session (1)[c]	3.9	−5.8 12.4	1.5 12.4	16.7[e] 13.0
Writing Expansiveness, end of session (1)[c]	1.4	−2.2 11.6	0.9 11.6	6.4 12.8
log Slow Writing (31)[a]	1.24	−0.02 0.48	0.09 0.48	0.01 0.54
Writing Intensity (Abstract Reasoning) (4)[c]	0.07	0.79 3.13	1.57 3.13	2.79 3.26
Writing Intensity (Free Association) (14)[c]	−0.43	0.14 1.77	0.50 1.77	0.79 1.87
Writing Intensity Composite[c]	−0.39	0.54 2.50	1.29 2.50	2.18 2.63

[a] Post-drug values of Tables 7, 9, 11, 13, and 15 through 20 were corrected by subtraction for the corresponding pre-drug values, except for measures (designated by the superscript [a]) lacking pre-drug values.
[b] The number following a measure is that assigned the test on which the measure is based.
[c] Expressive measure.
[e] Significant at 0.05 level.

Caffeine's acceleration of Writing Speed (1) was negligible (see Table 7). This may seem surprising, considering that caffeine has been found to allay impairment on prolonged tasks of a physically demanding nature (see, for example, Hollingworth [1912] and Seashore and Ivy [1953]). But Writing Speed's time limit was deliberately kept short to avoid marked fatigue. That Writing Speed was little (if at all) affected by drug-induced changes in fatigue is suggested by the finding of a non-significant interaction between drugs and halves of writing trials (though Ss' spontaneous comments indicated they felt more tired during the second half of each trial).

Note that beneficial caffeine effects have been observed for tapping, a task requiring speeded hand or finger movements and generally having time limits of the order of magnitude of those employed for Writing Speed; these beneficial effects have not usually been marked (Flory and Gilbert, 1943; Gilliland and Nelson, 1939; Hollingworth, 1912; Lehmann and Csank, 1957; Suhr, 1954; Thornton, Holck, and Smith, 1939). The most clear-cut findings are those of Hollingworth, who observed a steady increase in tapping rate with increasing dosage; the increase was about 5 per cent at Hollingworth's largest dose, 6 gr. caffeine. Too great an enhancement of function for brief tasks requiring rapid finger movements would consequently not be expected in the present study, whose average caffeine dose was slightly less than half of Hollingworth's maximum dose. This expectation conforms with the Writing Speed findings.[2]

Though impaired by the heavy alcohol dose, Writing Speed was not significantly affected by alcohol in the present study. Compare this finding with alcohol effects reported by other investigators. Travis and Dorsey (1929) reported a *speeding* of the latent time of the patellar tendon reflex. Are relatively primitive motor responses enhanced, due perhaps to a release of lower c.n.s. centers from the "inhibiting" effects of higher centers? Perhaps—see

---

[2] One of the more marked caffeine effects, about which there has been a surprising near-unanimity of report (Adler, Burkhardt, Ivy, and Atkinson, 1950; Gilliland and Nelson, 1939; Hollingworth, 1912; Hull, 1935; Lehmann and Csank, 1957; Stanley and Schlosberg, 1953; Suhr, 1954; Thornton *et al.*, 1939), is that of an *impairment* of hand steadiness. It is not clear, however, that such an impairment of steadiness would necessarily affect speed of writing.

Chapter 10. But deliberate control of one's writing speed is not quite so primitive a matter as reflex response.

More relevant to Writing Speed are the findings of an alcohol-induced impairment of tapping speed (impairment increased with dosage [Hollingworth, 1924]) and finger movement (see Jellinek and McFarland [1940][3]). Impairment of tapping and of other finger movements was mild, however; effects on speed of finger movement were among the mildest of a variety of alcohol effects observed by Hollingworth (1924) and by Miles (1924).

The lack of significant caffeine or alcohol effects for Writing Speed appears to be consistent with previous findings. It is therefore *unlikely that drug effects on tasks requiring much writing during brief intervals* (Free Association [14], for example) *were substantially*[4] *attributable to drug-induced changes in writing pace.*

A brief comment about Slow Writing (31). Observed values were transformed to logarithms before analysis. Treatment effects were found to be neglible. Firm conclusions cannot be drawn from this result, however, since Slow Writing proved to be an extremely unreliable measure. Lack of pre-drug values further inflated experimental error.

Ability to regulate writing pace (either by speeding it up or by slowing it down) was thus essentially unaffected either by alcohol or by caffeine under present conditions of experimentation.

Motor Control, Incidental Measures: Allport (1937) has emphasized the distinction between the involuntary, *expressive* aspects of behavior (as for example expansiveness or emphasis of bodily movements) and the volitional, *adaptive* aspects of behavior (exemplified by capacity tasks such as Writing Speed). Spontaneous as well as deliberate aspects of Ss' writing activity were observed in the present study. Ss' responses for Abstract

---

[3] Frequent reference is here made to the conclusions drawn by Jellinek and McFarland (1940) in their thoroughgoing critical review of the pre-1940 alcohol literature. The specific studies reviewed by Jellinek and McFarland, many of them based on but a single S, are much less frequently cited.

[4] Evidence for regression of Writing Speed on blood alcohol concentration is, however, presented in Chapter 8.

Reasoning (5) and for Free Association (14) were each rated for intensity of writing; these two indices were then combined to form the Writing Intensity Composite (see Table 3). Ss' writing was especially intensified by the heavy alcohol dose (see Table 7), suggesting that Ss, perhaps sensing an incipient reduction in motor control, might have "dug in" in an attempt to maintain or regain control. Had slightly more reliable writing intensity measures been available, the heavy alcohol dose effects might well have been significant.

More readily affected by drugs was Writing Expansiveness (1). With the finding of a significant drug × trials interaction (see Table 8), a separate analysis was conducted for each of the two trials on which Writing Expansiveness scores were based (see Table 7). Results for both trials indicate that Ss' handwriting became slightly more compact after ingestion of caffeine, and much more sprawling after ingestion of the heavy alcohol dose. The expansive effect of the heavy alcohol dose was quite powerful during the beginning-of-session writing trial, but had apparently worn off by session's end.

The increased writing spread associated with the heavy alcohol dose might be accounted for in several ways:

TABLE 8
ANALYSIS OF VARIANCE OF DRUG EFFECTS
ON WRITING EXPANSIVENESS (1)

Source of Variation	df	Mean Square	F	p
D (Drugs)	3	2,455.04	6.59	<0.005
B (Blocks)	13	481.34	1.29	NS[a]
D × B	37	372.29		
T (Trials)	1	1,063.14	9.24	<0.005
H (Halves of trials)	1	154.45	1.34	NS
T × H	1	85.02	0.74	NS
D × T	3	488.23	4.24	0.01
D × H	3	62.39	0.54	NS
D × T × H	3	44.81	0.39	NS
Residual	142	115.09		

[a] Failed to reach significance at the 0.05 level.

It has been suggested (Allport and Vernon, 1933; Eisenberg, 1937; Murphy, 1947) that expansiveness of movement is a function of S's self-assurance or of the pleasurableness of his state. The more expansive handwriting of the heavily alcoholized Ss might then be attributed to an elevation of mood and to an increase in self-confidence. But informal observation indicated that while these Ss were at times elated and at times self-confident, they were at least as often out of humor. Self-ratings for Satisfied with Myself and especially for Would Trust My Judgment (see Table 4) provide little support for the suggestion that the alcohol-induced writing spread was mediated by an alcohol-induced elevation of mood or increase in self-confidence.

As an alternative hypothesis, consider that alcohol effects on horizontal writing extent might perhaps have been a by-product of alcohol effects on fatigue.[5] This appears to be a fruitful suggestion, for those Ss whose writing expanded most following ingestion of the heavy alcohol dose reported the greatest post-drug increase in feelings of weariness. (Measures of writing spread obtained toward the beginning and end of sessions 1 and 3 were compared with self-ratings obtained at approximately the same times, i.e., toward the beginning and end of sessions 1 and 3. See Table 1.) But a closer look at the changes in Writing Expansiveness and in Tired and Weary self-ratings, following ingestion of the heavy alcohol dose, reveals that these changes were significantly correlated ($r = +0.57$, $df = 9$) only for session 3's final writing trial; changes in the two measures, following treatment, were essentially unrelated ($r = +0.10$, $df = 11$) for the beginning trial of session 3. Yet the results most relevant to the fatigue hypothesis are those for session 3's beginning trial, when blood alcohol concentration was near its peak value and when Writing Expansiveness was most markedly increased from its pre-drug values. The correlation between changes in writing spread and changes in feelings of weariness thus provides only minimal support for the fatigue hypothesis.

Further doubt about the fatigue hypothesis' adequacy is raised by the fact that Writing Expansiveness was primarily affected by

---

[5] Smith, K. U. Personal communication.

the heavy alcohol dose, while Tired and Weary self-ratings were almost exclusively affected by caffeine. The self-rating findings thus fail to corroborate the possibility that alcohol effects on horizontal writing extent were mediated by alcohol effects on fatigue.

Consider these further observations:

1. Writing pace may depend on a variety of factors, such as horizontal writing extent or the precision with which individual letters of the alphabet are formed. Pre-drug values for Writing Expansiveness and Writing Speed (1) were found to be reciprocally related. At the first writing trial of session 1, for example, these two measures correlated −0.52 ($p = 0.001$). Ss with the most expansive writing were thus observed to write least during a given time interval, and vice versa.

Despite their interrelation under non-drug conditions, speed and spread of writing differed in their response to drugs. Heavy alcohol dose effects, significant for Writing Expansiveness, fell far short of significance for Writing Speed.

This difference in alcohol effects suggests that *a task's deliberately adaptive aspects may be less susceptible to impairment by drugs than its involuntary expressive aspects.* For Ss had been explicitly and unmistakably directed to attend to the speed aspects of the writing task, to maximize writing output during a specified time interval. Yet Writing Speed remained stable after treatment, despite marked drug-induced changes in horizontal writing extent—an aspect of the task to which Ss had *not* been alerted. Though Ss' writing resources were perhaps impaired by the heavy alcohol dose, Ss seemed able to marshal their resources in order to attain the objectives set by the task. The action of the heavy alcohol dose appears to have been betrayed more readily by *incidental* features of the writing task, such as writing extent, than by such essential task features as writing speed. Detection of drug-induced impairment might perhaps be facilitated by noting shifts from Ss' spontaneous behavior patterns (compare Goldstein's "preferred behavior" [Goldstein, 1939]), as for example by observing non-capacity aspects of capacity tests.

2. Informal observations on non-capacity aspects of capacity tests have been made by investigators studying the effects of large

alcohol doses. Exner noted the *vehemence* with which S struck a lever used in studying reaction time (cited by Jellinek and McFarland). Goldberg (1943) noted the greater size and thickness of Ss' pencil markings, in a writing task. In the present study, Writing Intensity values tended to increase with alcohol dosage; this effect might have been more definite, had the Writing Intensity measures been more reliable. Alcohol thus appears to exaggerate centrifugal arm-and-hand movements, intensifying writing pressure as well as broadening the writing hand's horizontal sweep. (This generalization appears to be confirmed by the significant correlation of +0.38 [averaged over all four treatments] between treatment-induced changes in Writing Expansiveness and in the Writing Intensity Composite. But the correlation between these two measures was only +0.02 for the heavily alcoholized Ss, the very Ss in whom the most marked changes in motor control were observed.)

3. That Writing Expansiveness was more affected by the heavy alcohol dose than the Writing Intensity Composite is perhaps partly accounted for in terms of Smith, Harris, and Shideman's (1957) distinction between the "travel" and "manipulative" components of movement. In writing, for example, the hand travels along a frontal plane while engaged in manipulative movements. According to Smith *et al.*, the travel components of movement (of which Writing Expansiveness is but one index) are more affected by alcohol than are the manipulative components (of which the Writing Intensity Composite is presumably one index).

Supporting the findings of Smith *et al.* is Strongin and Winsor's (1933) observation that movements of the arm and hand (following a target in horizontal periodic motion) became less differentiated under alcohol's influence. The loss of differentiation in travel movements suggests an increase in inertia, the forearm being an inertial mass whose movements along a frontal plane are regulated by delicate nervous controls.

4. Strongin and Winsor further observed that arm movement became generalized into a *sway* of the entire body, with increasing alcoholization, as Ss attempted to follow the moving target. There was thus an increase in the inertial mass not only of the moving arm itself but of the entire moving system—the arm and

the rest of the body as well. Such a massive system has a sluggish response and readily overshoots its mark.

The cumulated evidence of increased motor inertia, of sway, and of reduced control over involuntary movements suggests that cerebellar functioning was impaired by the heavy alcohol dose. Gross signs of cerebellar disturbance are, of course, so frequently associated with marked alcohol intoxication that they tend to be regarded by the layman as a clear indication of drunkenness. One symptom of cerebellar disturbance, nystagmus, has been observed at blood alcohol concentrations as low as 70–80 mg. per cent (Goldberg, 1943), concentrations on the order of those obtained with the present heavy alcohol dose.

Granted the cerebellar impairment of the heavily alcoholized Ss, it would appear that the Writing Expansiveness findings represent a form of *hypermetria*, a symptom of cerebellar impairment characterized by overshooting the mark, due to reduced ability to adjust the force of the contraction needed to perform a given act. Possibly also involved were compensatory movements associated with vertigo, yet another symptom of cerebellar impairment.

Symptoms of cerebellar disturbance tend, furthermore, to go unobserved by those who suffer from them. Thus, Exner's subject (referred to above) was unaware of the vehemence with which he hit the lever. And Goldberg's Ss, whose performance on tests of cerebellar functioning was markedly impaired by alcohol, tended to be unaware of this impairment, and to be astonished on being shown objective evidence of impairment. Now one of the remarkable findings relating to Writing Expansiveness (this was noted informally) was Ss' failure to observe variations in horizontal writing extent and Ss' surprise on having these variations called to their attention on conclusion of the experiment. This observation suggests further that the alcohol-induced change in Writing Expansiveness reflected an alcohol-induced impairment of cerebellar functioning.

That cerebellar functioning was markedly impaired by alcohol is also suggested by the observations of Miles (1924), who found the "ataxiameter" (a device for measuring sway) to be a sensitive indicator of alcohol effects. Even more striking findings were obtained by Goldberg, who noted that logarithmic transformations of three measures of cerebellar functioning (the Romberg sign, both in its ordinary form and in modified form; and a quantitative version of the finger-finger test) increased linearly and significantly with blood alcohol concentration; x-intercepts for these regression lines were in the neighborhood of 50 to 60 mg. per cent. It is especially interesting that *these measures of cerebellar functioning were more markedly influenced by alcohol than were measures of other functions* studied by Goldberg.

Of the several hypotheses brought forward to account for the Writing Expansiveness findings, two—the cerebellar impairment and incidental task features hypotheses—seem most plausible. Neither explanation precludes the other. The greater wealth of evidence supporting the cerebellar impairment hypothesis, taken together with Goldberg's finding that alcohol's cerebellar effects can be revealed by a task's deliberately adaptive aspects, does however suggest that cerebellar impairment was the principal determinant of the heavily alcoholized Ss' more expanded writing.

Associative Productivity: Drug effects on associative productivity were marked. Although none of the findings for the first half-dozen measures listed in Table 9 exceeded the level of significance, there was a striking similarity in the pattern of findings for all but one of these measures (see Fig. 2).

A clear appreciation of the drug effects on associative productivity can be obtained by analytical as well as by graphical means. Free Association (14) and Word Fluency (15) were combined (each measure was assigned the same scale factor) to form a Word Association Composite. This composite reveals the drug effects more clearly than do its component measures. Also formed was the more heterogeneous Associative Productivity Composite, combining all five association measures represented in Fig. 2, including Free Association and Word Fluency; see Table 3 for the exact make-up of this composite.

It is evident, both from Table 9 and from Fig. 2, that caffeine enhanced associative productivity. The caffeine effect, clearly significant for both composite measures, closely approached significance for Free Association, Word Fluency, and Consequences: Low-quality (7). That caffeine affected least the measures involving the greatest constraint on association suggests that the

Intellectual Functioning 57

quantity of energy available for intellectual work is more readily affected by caffeine than are the psychological or central nervous *structures* determining the paths along which this energy is directed. But see the section: *Fixed associations, overlearned,* below.

Rate and abundance of association were less markedly affected by alcohol than by caffeine. The measures entering into the Associative Productivity Composite, while hardly affected by the heavy alcohol dose, were fairly consistently enhanced by the small alcohol dose.

The increase in associative productivity with the small alcohol dose, though not statistically significant, appears to have been more than an isolated, chance finding. For Horace wrote: "What wonders does not wine! . . . Whom has not a cheerful glass made eloquent!" (cited by Goodman and Gilman [1955]).[6] *Experimental* evidence of association facilitated by "depressant" drugs is also available: Von Felsinger, Lasagna, and Beecher (1953) observed a significant increase in associations to nonsense syllables, following administration of a hypnotic dose of a barbiturate.

Of special interest is the excellent study of Hartocollis and Johnson (1956), who administered an alcohol dose (37.9 cc. absolute per 100 lbs. of body weight) somewhat larger than the present heavy alcohol dose. The alcohol and control doses used by Hartocollis and Johnson were essentially identical in composition to the present alcohol and control doses. These investigators observed a significant alcohol-induced reduction of associative productivity for measures of relatively unconstrained association that resembled Free Association and Word Fluency. Measures of verbal association requiring greater constraint were similarly, but less definitely, affected by alcohol.

Hartocollis and Johnson's findings appear to be an extrapolation of the present findings. These combined findings, taken together with the impressions noted informally by Loomis and West (1958) —who reported that Ss were more talkative at blood alcohol concentrations below 100 mg. per cent than at concentrations above

---

[6] Or compare Roget, on the effects of nitrous oxide: "My ideas succeeded one another with extreme rapidity, thoughts rushed like a torrent through my mind" (cited by Freedman [1960]).

TABLE 9
Treatment Effects on Association

| Measure | Placebo Mean | Difference Between Means of Active Treatment and Placebo |||
		CAFFEINE	ALCOHOL (15.7 ML.-/M.²)	ALCOHOL (31.4 ML.-/M.²)
ASSOCIATIVE PRODUCTIVITY:				
Free Association (14)	−0.07	7.00 7.01	3.86 7.01	1.00 7.38
Word Fluency (15)	2.36	6.00 6.67	0.36 6.76	−0.14 6.95
Controlled Associations: Total (10)	3.86	2.57 4.94	1.50 4.94	−0.29 4.94
Controlled Associations: Synonyms (10)	4.86	0.86 4.60	0.36 4.60	−1.07 4.60
Consequences: Low-quality (7)	−0.79	2.93 3.18	0.29 3.18	−2.29 3.18
log ([Plot Titles: Low-quality] + 2) (11)	0.08	0.02 0.18	−0.01 0.18	0.10 0.19
Word Association Composite	0.57	3.25 * 2.72	1.05 2.72	0.21 2.86
Associative Productivity Composite	0.052	0.097* 0.092	0.032 0.092	−0.005 0.092
log Sentence Completion: Word Count (28)[a,c]	1.62	0.21 * 0.20	0.18 0.20	0.08 0.22
HIGH-QUALITY ASSOCIATIONS:				
Consequences: Remoteness (7)	2.07	−1.64 2.94	1.14 2.94	−0.29 2.94
Plot Titles: Cleverness (11)	−0.43	0.79 2.57	−0.36 2.57	−0.43 2.70
FIXED ASSOCIATIONS, OVERLEARNED:				
Stroop Color-naming Time (18)	−1.79	−1.43 4.38	2.29 4.38	2.22 4.90
Addition, Correct (8)	−1.57	4.57 4.69	4.21 4.69	−1.79 4.69
Addition, Incorrect (8)	0.50	−0.64 2.57	−1.71 2.57	0.29 2.57

[a] Pre-drug values lacking.
[c] Expressive measure.
* Significant at 0.05 level.

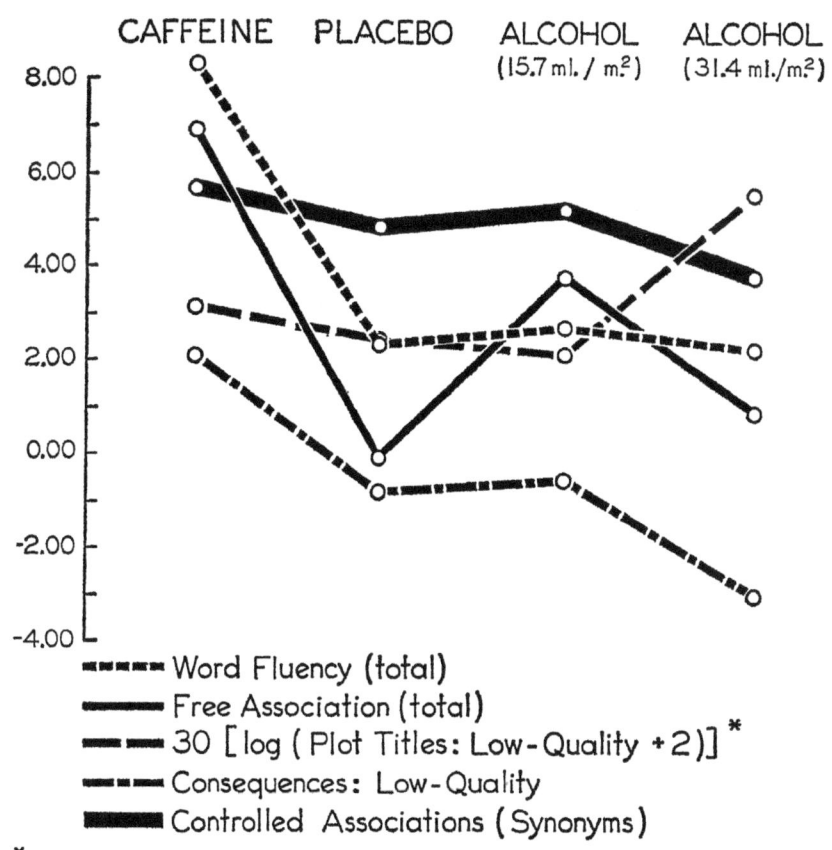

CAFFEINE   PLACEBO   ALCOHOL   ALCOHOL
                    (15.7 ml./m²) (31.4 ml./m²)

▰▰▰▰ Word Fluency (total)
━━━━ Free Association (total)
━ ━ ━ 30 [log (Plot Titles: Low-Quality + 2)] *
━ ▪ ━ Consequences: Low-Quality
▰▰▰▰ Controlled Associations (Synonyms)

*"30" is an arbitrary factor used for this figure only

Fig. 2. Associative productivity measures: treatment means.

100 mg. per cent—suggest the following *biphasic hypothesis:* Rate and abundance of association increase with small doses of alcohol, say up to 25 cc. absolute; this effect is gradually reversed, increased associative productivity giving way to decreased productivity at alcohol doses of the order of 50 cc. absolute.

Findings for Sentence Completion: Word Count (28)—this measure was transformed to logarithms—resembled those for the Associative Productivity Composite (see Table 9): wordiness was increased both for caffeine and for the small alcohol dose but was hardly affected for the heavy alcohol dose. Caffeine's effect on Word Count was significant, whereas the small alcohol dose effect was of borderline significance. These findings lend support to the biphasic hypothesis, particularly since the small alcohol dose had a more marked effect on Word Count than on measures entering into the Associative Productivity Composite.

The level of significance attained by the Word Count findings is remarkable, in view of the imprecision and the absence of predrug values for this measure. Two equally plausible hypotheses are suggested to account for Word Count's special sensitivity to drug effects, relative to that of the remaining measures of verbal associative productivity:[7]

1. Conveyed by the Sentence Completion Test instructions was an emphasis on the "content" or meaning of the responses (Nash, 1958). In contrast with the capacity measures of the Associative Productivity Composite (for which Ss were encouraged to maximize quantity of associations), the Sentence Completion Test did not encourage Ss to respond in conformity with prescribed standards of intellectual merit. If there was any tendency to constrain the intellectual form of Ss' responses, it arose from the instruction to complete sentences rapidly, which instruction might indirectly have encouraged a reduction (rather than an increase!) in sentence length. Word Count is then an "incidental," non-capacity measure, applied to the Sentence Completion responses without Ss' foreknowledge; compare the section: *Motor control, incidental measures*, above. Here, as for Writing Expansiveness, drug effects may more readily have been revealed by incidental measures reflecting involuntary aspects of Ss' behavior than by capacity measures directly indicating the extent to which task objectives were attained. See the section: *The role of attention*, Chapter 11.

2. The drugs' primary associative actions may more nearly have been reflected by Word Count than by the measures entering into the Associative Productivity Composite. Caffeine and the

---

[7] Given the lack of pre-drug values for Word Count and given the somewhat higher Vocabulary scores of Ss assigned caffeine and the small alcohol dose, might the Word Count findings simply have reflected the intellectual superiority of the caffeine and small alcohol dose Ss (see the section: *Subjects*, Chapter 4)? The absence of correlation between *log* Word Count and Vocabulary fails to provide support for this suggestion.

small alcohol dose may have acted primarily to increase "wordiness" of communication; findings for the Associative Productivity Composite, some of whose component measures required Ss to maximize quantity of ideas rather than quantity of words, may not then have provided a direct reflection of the drugs' essential associative effect.

"Quality" of Associations: The work of Guilford and associates (see Guilford et al. [1957] and especially Kettner et al. [1959]) suggests that High-quality scores for Consequences (7) (i.e., Remoteness) and for Plot Titles (11) (i.e., Cleverness) are indices of originality, while Low-quality scores for these two tests reflect "ideational fluency" (see the section: *Associative productivity*, above). Combination of Low- and High-quality scores for both tests into a Consequences-Plot Titles Composite reveals a significant interaction between drugs, tests, and "qualities" (see Tables 3 and 10). The implications of this finding are not readily apparent.

Fixed Associations, Overlearned: The pattern of findings for Stroop Color-naming Time (18) (see Table 9) departed from that for the verbal associative productivity measures. Though accel-

TABLE 10

ANALYSIS OF VARIANCE OF DRUG EFFECTS ON THE
CONSEQUENCES–PLOT TITLES COMPOSITE

Source of Variation	df	Mean Square	F	p
D (Drugs)	3	0.0162	0.52	NS[a]
B (Blocks)	13	0.0144	0.46	NS
D × B	39	0.0310		
T (Tests)	1	0.0046	0.15	NS
Q (Qualities: High vs. Low)	1	0.0027	0.09	NS
T × Q	1	0.9439	31.05	<0.005
D × T	3	0.0678	2.23	NS
D × Q	3	0.0361	1.19	NS
D × T × Q	3	0.1290	4.24	0.01
Residual	152	0.0304		

NOTE: See Table 3 for the make-up of this composite.
[a] Failed to reach significance at the 0.05 level.

erated by caffeine, color-naming was retarded by *both* alcohol doses. While not statistically significant, these findings agree with other published findings. Thus, Hollingworth (1912) reported an acceleration of color-naming with caffeine. Studying alcohol effects, Hollingworth (1924) observed that color-naming was retarded with increasing alcohol dosage. Somewhat similar alcohol findings were obtained by McFarland and Barach (1936) and by Carlson, Kleitman, Muehlberger, McLean, Gulliksen, and Carlson (1934).[8]

The adding task also involved associations rigidly constrained by the task and repeated to the point of being almost mechanical. Addition, Correct (8) was found to be enhanced by caffeine; this finding was of borderline significance. A (non-significant) increase in adding rate was also observed by Gilliland and Nelson (1939), who studied the effects of coffee containing quantities of caffeine similar to those administered in the present study. The present findings of a caffeine-induced enhancement of adding rate resemble the caffeine findings for other association measures. Since adding involves constraint of association, the addition findings call into question the hypothesis developed above (see the section: *Associative productivity*) that the amounts of energy available for intellectual work are more readily affected by caffeine than are the psychological or central nervous structures guiding this energy's utilization.

Alcohol effects on adding rate are of special interest. The enhancement of Addition, Correct by the small alcohol dose approached significance, while the effect of the larger dose was negligible. These findings resemble those for the verbal associative productivity measures, lending further support to the biphasic hypothesis formulated above (see the section: *Associative productivity*).

Equally striking is the (non-significant) finding that errors in addition were *decreased*, rather than increased, by the small alcohol dose. The ability of the lightly alcoholized Ss to add more rapidly thus indicates an actual improvement of intellectual ca-

---

[8] That the small alcohol dose enhanced verbal associative productivity but not color-naming might perhaps have been due to an alcohol-induced disturbance of color vision.

pacity; it was no mere by-product of lowered standards of performance.

The results reported in the immediately preceding paragraphs were based on brief addition tasks. But there have been many studies in which Ss continued to add for lengthy periods. Barmack's (1940) results were especially notable. He required Ss to add pairs of six-place numbers, the top number being always the same. Barmack found that adding rate was usually unaffected, during the first fifteen minutes of a continuous addition task, by a dose of caffeine only slightly smaller than that employed in the present study. But caffeine did *allay* impairment of adding rate by boredom and/or fatigue, which grew in intensity during the two hours of continuous addition. (Barmack [1938] reported similar results for amphetamine and other c.n.s. stimulants.)

The discrepancy between Barmack's finding of an initial adding rate unaffected by caffeine and the present finding of a caffeine-induced enhancement of adding rate appears attributable to differences between the tasks employed in the two studies. Barmack's adding task was intended to be boring; that Barmack's Ss did indeed experience much ennui and/or fatigue while working at their lengthy adding task is evident from their report that, while under the influence of placebo, between 10 and 20 per cent of the time allotted to the adding task had been spent *daydreaming*. (See the discussion of self-rating findings.) Contrast the task attitude of Barmack's Ss with that of the present Ss, who approached their five-minute adding task as if primed for a short-distance run and proceeded to complete the task, once the starting signal had sounded, without delay or distraction.

The ability to add rapidly thus appears to be enhanced by caffeine. But when Ss' motivation is insufficient to satisfy boundary conditions (cf. Nash, 1959) for the measurement of adding capacity (as when an experimental situation fails to arouse Ss' interest and to engage their full attention), caffeine effects on adding capacity may be confounded with and obscured by caffeine effects on attitude.

Visual Thinking: Of the "visual thinking" measures listed in Table 11, all but Stroop Reading Time (18) were combined to form composite measures. In the absence of a significant interaction between drugs and parts of the test, Parts I and II of Clerical Speed and Accuracy (21) were combined (each measure was assigned the same scale factor) to form a Perceptual Speed Composite. Also combined were Mutilated Words (25) and Street Gestalt Completion (16) (here too each measure was assigned the same scale factor); the resulting Closure 1 Composite reflects the ease of re-closure of meaningful visual patterns which had previously been disrupted (cf. Thurstone, 1944; Thurstone, 1949). Drug findings for the Closure 1 Composite are meaningful, despite a significant drugs × tests interaction (see Table 12), since the level of significance for the drug effects greatly exceeded that for interaction.

TABLE 11
TREATMENT EFFECTS ON VISUAL THINKING

Measure	Placebo Mean	CAFFEINE	ALCOHOL (15.7 ML.-/M.^{2})	ALCOHOL (31.4 ML.-/M.^{2})
Clerical Speed and Accuracy: Part I (21)	4.71	−0.57 6.97	−3.21 6.97	−12.71* 7.80
Clerical Speed and Accuracy: Part II (21)	0.79	2.79 7.02	1.07 7.02	−4.79 7.67
Perceptual Speed Composite (21)	2.75	1.11 5.82	−1.07 5.82	−8.75* 6.51
Mutilated Words (25)	1.39	0.04 2.87	−0.04 2.87	−3.89* 3.33
Street Gestalt Completion (16)	−0.14	0.36 2.05	−0.64 2.05	−0.50 2.30
Gottschaldt Figures (22)	−1.71	2.36 4.88	0.07 4.88	−1.14 5.65
Closure 1 Composite	0.63	0.20 1.70	−0.34 1.70	−2.20* 1.84
Visual Thinking Composite	3.7	5.0 11.5	−2.7 11.5	−21.2 * 12.5
Stroop Reading Time (18)	0.93	−0.22 2.71	−0.14 2.71	2.50 3.03

*Significant at 0.05 level.

TABLE 12

ANALYSIS OF VARIANCE OF DRUG EFFECTS ON CLOSURE 1 TESTS
(MUTILATED WORDS [25] AND STREET GESTALT COMPLETION [16])

Source of Variation	df	Mean Square	F	p
D (Drugs)	3	33.69	5.13	<0.005
B (Blocks)	13	7.54	1.15	NS[a]
D × B	36	6.57		
T (Tests)	1	16.13	2.25	NS
D × T	3	22.33	3.11	0.05
Residual	43	7.17		

[a] Failed to reach significance at the 0.05 level.

A more inclusive, but also more heterogenerous, Visual Thinking Composite was formed by combining the Perceptual Speed Composite with three other visual thinking measures (see Table 3).

The visual thinking measures were rather consistently enhanced by caffeine (negative values of Stroop Reading Time indicate enhancement), but the results in no instance approached significance. The work of previous investigators (Adler, Burkhardt, Ivy, and Atkinson, 1950; Flory and Gilbert, 1943; Hollingworth, 1912; Lehmann and Csank, 1957) tends in general to support the present findings. Taken together, these various results suggest that visual thinking is enhanced *to a very minor extent* by caffeine.

Even more negligible were the (negative) effects of the small alcohol dose on visual thinking. In striking contrast was the marked disturbance of function produced by the heavy dose of alcohol. *Each* of the visual thinking measures was impaired by the heavy alcohol dose; this impairment was significant for Mutilated Words and for the three composites. The *extreme* impairment of the Visual Thinking Composite by the alcohol dose is the *most marked* of the drug effects observed in the present study.

Many other investigations of alcohol effects on processes related to vision have been conducted. Goldberg's (1943) findings are particularly interesting. He reported a significant regression of brightness for fusion on blood alcohol concentration. Regression was linear when brightness for fusion was transformed to logarithms, drug effects becoming noticeable (for moderate drinkers) at blood alcohol concentrations of the order of 50 mg. per cent. These, as well as related findings of other investigators, were attributed by Goldberg to an impaired functioning of the retina and of the optic nerve.

Findings relating to visual acuity have been less clear-cut. Marquis, Kelly, Miller, Gerard, and Rapoport (1957) failed to observe a change in visual acuity, for either near vision or far vision; but the alcohol dose employed by these investigators was small. Little change in visual acuity was observed by Moeller and Becker or by Colson, even though large quantities of alcohol were employed (results cited by Goldberg). On the other hand, Jellinek and McFarland interpreted the evidence obtained by Ridge, by Miles, and by Busch as indicating some impairment of visual acuity with small to moderate quantities of alcohol. Impaired visual acuity was also reported by Powell (cited by Goldberg). Most decisive are the findings of Newman and Fletcher (1941), who reported that vision became less acute as alcohol concentration increased; but a really marked impairment of acuity (such as a drop from 20/20 to 20/45, on the Snellen chart) rarely occurred except when blood alcohol concentrations exceeded 100 mg. per cent. These findings suggest that visual acuity was impaired by the present heavy alcohol dose but was hardly affected by the small alcohol dose.

Deleterious alcohol effects on eye movements have also been observed. Newman and Fletcher found eye co-ordination (studied under non-speeded conditions) to be disturbed by large quantities of alcohol. Eye co-ordination grew worse with increasing blood alcohol concentration, but deterioriation of function was slightly less marked here than for visual acuity. Dodge and Benedict (1915), as well as Miles (1924), observed a slight slowing of eye movements after moderate doses of alcohol.

Findings of other investigators relating to alcohol effects on cancellation of letters have been quite clear. Impairment of cancellation by moderate amounts of alcohol was observed in several small studies cited by Jellinek and McFarland. The excellent studies of Goldberg (1943) and of Takala, Siro, and Toivainen (1958), in which large alcohol doses were employed, provide even more definite evidence of impaired cancellation. Goldberg reported a significant regression of cancellation time on blood alcohol concentration; regression was linear when cancellation time was transformed to logarithms, drug effects becoming noticeable (for moderate drinkers) at blood alcohol concentrations of the order of 40 to 70 mg. per cent.

Taken together with the findings of others, the present findings provide conclusive evidence of a *very marked* impairment of visual function with large quantities of alcohol. With small quantities of alcohol, on the other hand, impairment is mild, perhaps negligible. The work of Newman and Fletcher (1941) indicated that serious disturbance of simple visual functions begins to set in only after attainment of blood alcohol levels of the order of those found for the heavily dosed Ss of the present study. Goldberg's findings were even more specific: deleterious effects on visual functioning became noticeable at blood alcohol concentrations between 40 to 70 mg. per cent for moderate drinkers, and between 10 to 50 mg. per cent for abstainers. *Peak* blood alcohol concentrations attained by the present study's mildly dosed group (which included both moderate drinkers and near-abstainers) were therefore *at* the threshold of impairment of visual function; *average* blood alcohol concentrations for this group tended to fall *below* the threshold of impairment. These observations, plus Goldberg's finding of a *logarithmic* increase in impairment with increasing blood alcohol concentration, account both for the small alcohol dose's negligible effect and for the large alcohol dose's clearly disturbing effect on the Visual Thinking Composite.

No clear ordering of visual measures according to degree of alcohol susceptibility is indicated by the findings reported above. Newman and Fletcher reported differential alcohol effects on measures of visual functioning; especially pertinent is their finding that visual acuity was more disturbed by alcohol than eye coordination. In the present study, the Perceptual Speed Composite was more impaired by alcohol than the Closure 1 Composite; but this comparison is weakened by the lack of precision of Street Gestalt Completion (16), one of the two components of the Closure 1 Composite. That alcohol effects for cancellation exceeded those for tachistoscopic recognition of meaningful visual patterns was suggested by Jellinek and McFarland; but the tachistoscopic findings were based on quite a small sample of Ss. No appreciable difference was observed between alcohol effects on cancellation time and on brightness for fusion (Goldberg), with respect either to slope or to intercept of the regression line.

Although successful performance for the present visual thinking measures depends on factors such as discrimination of small differences in visual pattern, perception of meaningful patterns on the basis of minimal cues, co-ordination of eye movements, and eye-hand co-ordination, impairment of visual thinking by alcohol cannot be attributed to any single factor. Each of these factors appears to be affected to some extent by large quantities of alcohol.

Visual-motor Co-ordination: The activities called for by Digit Symbol (17) and by Clerical Speed and Accuracy (21) have several features in common: the eye must glance back and forth in a swift and controlled manner; small patterns must be discriminated visually; the writing hand must move rapidly along the test protocol. But it is just this set of activities that is here designated visual-motor (more specifically, eye-hand) co-ordination.

Results for the first two Digit Symbol trials of session 4 were summed (after subtraction of the corresponding results for ses-

TABLE 13
Treatment Effects on Visual-Motor Co-ordination

Measure	Placebo Mean	Difference Between Means of Active Treatment and Placebo		
		CAFFEINE	ALCOHOL (15.7 ML.-/M.^{2})	ALCOHOL (31.4 ML.-/M.^{2})
Digit Symbol Composite 1 (17) ..	2.16	0.09 1.72	−0.43 1.72	−2.93* 1.93
Digit Symbol, trial 1, first 30″ ..	0.93	−0.50	0.14	−4.43
Digit Symbol, trial 1, second 30″ .	2.93	0.71	−0.79	−2.63
Digit Symbol, trial 2, first 30″ ..	1.64	−0.64	−0.43	−1.54
Digit Symbol, trial 2, second 30″ .	3.14	0.79	−0.64	−4.34
Digit Symbol Composite 2 (17)ᵃ .	23.73	0.39 2.77	−0.25 2.77	−3.13* 3.10

ᵃ Pre-drug values lacking.
* Significant at 0.05 level.

TABLE 14

ANALYSIS OF VARIANCE OF DRUG EFFECTS ON DIGIT SYMBOL COMPOSITE 1 (17)

Source of Variation	df	Mean Square	F	p
D (Drugs)	3	113.84	8.43	<0.005
B (Blocks)	13	14.59	1.08	NS[a]
D × B	35	13.50		
T (Trials)	1	12.54	2.64	NS
H (Halves of trials)	1	166.29	35.01	<0.005
T × H	1	15.54	3.27	NS
D × T	3	0.60	0.13	NS
D × H	3	13.09	2.76	0.05
D × T × H	3	10.74	2.26	NS
Residual	140	4.75		

[a] Failed to reach significance at the 0.05 level.

sion 2) to obtain Digit Symbol Composite 1; findings for the third and fourth trials of session 4 were omitted from Digit Symbol Composite 1. Digit Symbol Composite 2 was formed by summing the results for all four Digit Symbol trials of session 4; pre-drug values were omitted from Composite 2. Average results per half-trial are presented in Table 13, for both composites.

Caffeine affected neither Digit Symbol composite. This finding seems to contrast with that reported by Lehmann and Csank (1957), who noted significant improvement on a digit symbol test, with caffeine; but this improvement, of the order of magnitude of that manifested by their placebo Ss, was apparently due to practice rather than to caffeine. Consistent with the absence of caffeine effects for Digit Symbol is the lack of clear-cut caffeine findings for visual-choice reaction time (see Hollingworth [1912]; Stanley and Schlosberg [1953]; Thornton et al. [1939]) and for pursuitmeter performance (Adler et al., 1950; Suhr, 1954)—this latter measure involves both visual-motor co-ordination and learning, but of a more complex sort than that required for Digit Symbol. Tasks requiring complex visual-motor co-ordination and involving an element of learning appear to be little affected by caffeine.

Digit Symbol Composite 1 was markedly influenced by the experimental treatments (see Table 14), despite the absence of caffeine effects. Drug effects for the composite as a whole are of interest, notwithstanding the significant interaction between drugs and halves of trials, as the interaction was overshadowed by drug effects for the composite as a whole. The significant drug findings for Composite 1 are accounted for by the very deleterious effects of the heavy alcohol dose; Composite 1 was hardly affected by the small alcohol dose. Similar findings were obtained for Composite 2. The alcohol results are in accord with previous findings: Hollingworth (1924) found digit symbol performance little impaired by relatively small quantities of alcohol; McFarland and Barach (1936) found somewhat greater impairment with large quantities of alcohol.

How may one account for the heavy alcohol dose's markedly deleterious effects on Digit Symbol performance? These effects, like those for visual thinking, are attributable to a multiplicity of factors. That visual acuity, eye co-ordination, and eye-hand coordination are implicated in the Digit Symbol findings seems likely, given the points of resemblance between the Digit Symbol and Clerical Speed and Accuracy tests. Digit Symbol involves still further factors, such as the recall of arbitrary associations between stimuli (see the section: *Recall*, below).

Perhaps also relevant to the Digit Symbol findings is a subjective impression gained by Loomis and West (1958) while studying alcohol effects on a simulated automobile driving task. Loomis and West were of the opinion that large quantities of alcohol disrupted Ss' ability to divide their attention effectively between competing sets of visual stimuli. A reduced effectiveness in shifting attention from one locale to another may have played a part in the alcohol-induced impairment of Digit Symbol performance.

Recall: The first five measures listed in Table 15 each require S to memorize and reproduce immediately material that has been presented orally. Despite the virtual absence of significant findings for these recall measures, it is clear that recall was markedly affected by the experimental treatments. Performance was enhanced by caffeine in four (out of five) instances; see Fig. 3. The impression gained from Fig. 3 is confirmed by combining recall measures, as in the Associate Learning Composite (made up of Hard and Easy Associations, each assigned the same scale factor). Especially striking are the findings for the Immediate, Intentional Recall Composite (a combination of the five recall measures referred to above; see Table 3), which indicate that the immediate recall of auditory material, intentionally committed to memory, was significantly improved by caffeine (see Table 15). The recall results of previous caffeine studies (Lehmann and Csank [1957] observed a non-significant enhancement of digits forward, while Hull [1935] reported a non-significant impairment of nonsense-syllable learning rate) are meager and hardly modify the picture that emerges from the present findings.

Recall was decidedly disturbed by the heavy alcohol dose. Four of the five recall measures included in Fig. 3 were impaired, and impairment of the Immediate, Intentional Recall Composite was significant. This composite was not affected by the small alcohol dose.

The data of Davis, Gibbs, Davis, Jetter, and Trowbridge (1941) suggest a relation between immediate memory and blood alcohol concentration. Note, however, that these data were obtained with quantities of alcohol exceeding those administered in the present study. Further evidence of an alcohol-induced impairment of immediate, intentional recall has been presented by Jellinek and McFarland. Reviewing the few small-scale studies conducted before 1940, Jellinek and McFarland concluded that all types of memory (e.g., paired-associate learning, rote memory, memory for verse) showed great impairment, even with small alcohol doses. This conclusion is supported by the present findings for the large alcohol dose but not for the small alcohol dose.

In contrast to intentional recall, incidental recall appears to have been unaffected by the active treatments. Well-defined trends were lacking for the two incidental recall measures, Estimation of Digit Symbol, Corrected (17) and Incidental Learning (33) (see Table 15). The absence of drug effects for incidental recall suggests that drug effects on intentional recall were not mediated by drug effects on retention.

But the evidence provided by the two incidental recall measures is of only limited value. These measures were fewer and less precise than the intentional recall measures. Estimates of error for incidental recall were increased because of the lack of pre-drug values. Incidental Learning was, moreover, always measured toward the very end of session 4, when the drug levels of the blood had generally declined below the levels existing for the great majority of tests, which had been administered in random order.

TABLE 15
TREATMENT EFFECTS ON RECALL

Measure	Placebo Mean	CAFFEINE	ALCOHOL (15.7 ML./M.^{2})	ALCOHOL (31.4 ML./M.^{2})
Digits Forward (3 trials) (2)	1.00	1.00 / 2.07	0.64 / 2.07	−0.93 / 2.12
Associate Learning: Hard Associations (13)	−0.79	1.36 / 1.75	−0.29 / 1.75	−0.64 / 1.85
Associate Learning: Easy Associations (13)	−1.50	2.14* / 1.60	0.50 / 1.60	0.00 / 1.69
Sentence Memory (5)	1.14	−0.46 / 1.27	0.43 / 1.27	−0.32 / 1.33
Story Memory: Immediate Recall (20)	−0.36	1.61 / 3.06	−2.22 / 3.06	−3.00 / 3.42
Associate Learning Composite	−0.57	0.88* / 0.66	0.05 / 0.66	−0.16 / 0.70
Immediate, Intentional Recall Composite	0.00	2.97* / 2.34	−0.03 / 2.34	−2.40* / 2.40
Story Memory: Delayed Recall (23)[a]	−1.07	0.14 / 1.53	−0.79 / 1.53	−0.04 / 1.78
Incidental Learning (33)[a]	4.82	−0.12 / 1.96	−0.05 / 1.96	0.47 / 2.36
Estimation of Digit Symbol, Corrected (17)[a, b]	−1.79	2.15	−0.96	3.54

[a] Pre-drug values lacking.
[b] Differences between means evaluated by H-test.
* Significant at 0.05 level.

Story Memory findings have a greater bearing on the question of drug effects on retention. Compare drug effects on the Immediate Recall (20) of stories and on their Delayed Recall (23). Although falling short of significance, results for Immediate Recall were nevertheless among the most clearly defined of the immediate, intentional recall findings. Drug effects on Delayed Recall were, by contrast, quite negligible (see Table 15). Although the rehearsal associated with an immediate recall tends to hinder the detection of treatment effects on later recalls, the contrasting findings for Immediate and Delayed Recall nevertheless lend some support to the idea that retention was little, if at all, affected by the active treatments (retention is involved in Delayed Recall as in Immediate Recall). The implications, for retention, of the immediate-delayed recall contrast are perhaps clearer than the impli-

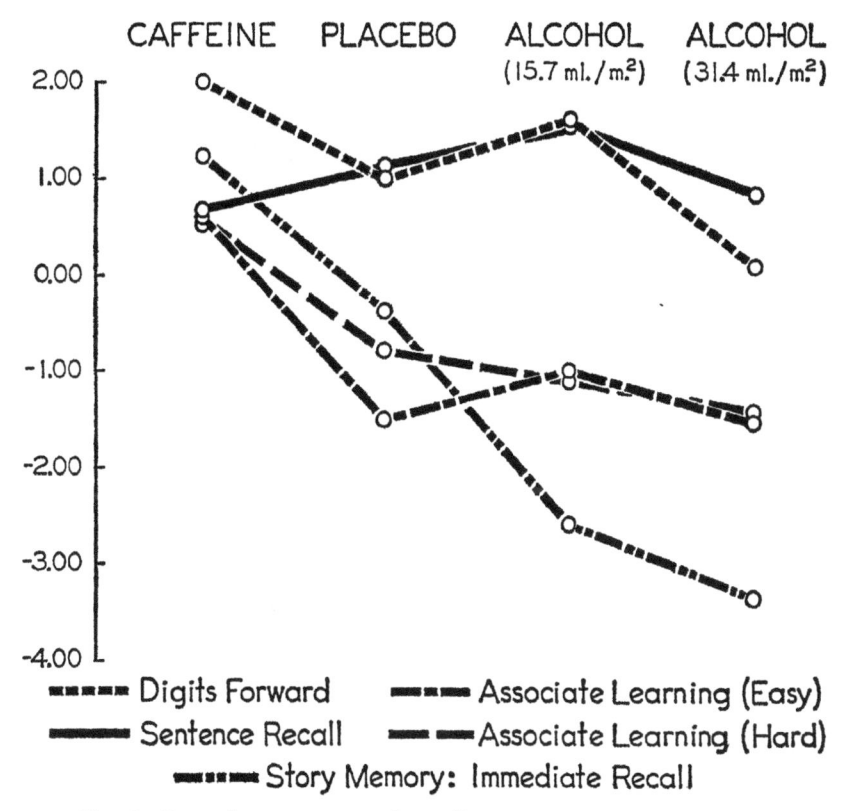

Fig. 3. Immediate, intentional recall measures: treatment means.

cations of the intentional-incidental recall contrast, for Immediate and Delayed Recall each contained corrections for baseline values, and they were measures of a similar order of precision.

Drug effects on immediate, intentional recall, not likely mediated by drug-induced changes in retentive capacity, appear to have been induced while the "memory trace" was being formed, before the trace had a chance to stabilize. Might the drugs have altered auditory receptivity, thereby affecting immediate, intentional recall? Might caffeine have enhanced or might the heavy alcohol dose have impaired the capacity to *receive* the rapid influx of auditory information provided by each of the tests of immediate, intentional recall? A negative answer is suggested by the negative results for Incidental Learning, which also required the reception of a rapid influx of auditory information; but the Incidental Learning findings were of too limited a value to support any firm conclusions.[9]

That changes in auditory receptivity were responsible for the drug effects on immediate, intentional recall seems most doubtful, however, when one considers the manner of presentation of the auditory stimuli used in the recall tests. These stimuli were presented to Ss through earphones. Each set of earphones had its own volume control. Each S was encouraged to adjust his volume control until the sound intensity was comfortable for him, until he experienced the sound level as being neither too loud nor too soft. Had the drugs in question affected auditory acuity, Ss would have been able to alter the sound volume to compensate for acuity changes. It is clear, therefore, that unlike the drug effects on visual thinking, which were apparently mediated in part by changes in visual acuity, the drug effects on immediate, intentional recall were not likely effected by changes in auditory acuity.

More likely than the "retention" and the "auditory acuity" hypotheses is the hypothesis that drug effects on immediate, intentional recall were brought about by changes in Ss' capacity to take full advantage of the rapid influx of information, to organize this information so that it might best be assimilated and formed

---

[9] While an alcohol-induced impairment of auditory acuity has been reported, the existence of such an effect has not been clearly demonstrated (Jellinek and McFarland; McFarland, 1953).

into a stable memory trace. Consider the evidence. Story Memory: Immediate Recall required active efforts to organize incoming auditory information; Story Memory: Delayed Recall called for no further active efforts at organization (Ss were not forewarned of the delayed recall). It was Immediate Recall rather than Delayed Recall that responded to the active treatments. Furthermore, active efforts at organizing incoming information were called for by *all* of the immediate, intentional recall measures, but by *none* of the incidental recall measures. It was immediate, intentional recall rather than incidental recall that was affected by drugs. The findings for these contrasting sets of recall measures are therefore consonant with the hypothesis that drug-induced changes in immediate, intentional (auditory) recall were mediated by drug-induced changes in Ss' ability to "lay hold of" and to organize the instreaming auditory information. Also see the sections: *Learning* and *Flexibility of thinking*, below.

Learning: Investigated in this study were drug effects on two aspects of learning: memorizing (see the section *Recall*, above) and "learning to learn." Information about "learning to learn"—an important factor underlying improvement from trial to trial in tests consisting of sequences of trials—was provided by Digits Forward (2), Digits Backward (3), Associate Learning (13), and Digit Symbol (17).

Digits Forward and Digits Backward were two tests of almost identical structure, each containing three trials per level of difficulty. An analysis for these two tests *combined* revealed that neither the drugs × trials interaction nor the drugs × tests × trials interaction was significant. Similar findings were obtained for the Associate Learning test, whose Hard and Easy Associations subtests were also of almost identical structure. The drugs × trials and the drugs × trials × trial-halves interactions for Digit Symbol Composites 1[10] (see Table 14) and 2 were also lacking

---

[10] There was, however, a significant interaction between drugs and halves of trials for Digit Symbol Composite 1 (see Table 14), reflecting the fact that caffeine had a more beneficial effect on second halves than on first halves of the Digit Symbol trials. This selective caffeine effect may have been due to an improved recall of arbitrary associations between digits and symbols (see the section: *Recall*, above), obviating the need to glance back and forth between digit and symbol; or it may have been due to a forestalling of fatigue (see Chapter 5).

in significance. Learning to learn seems hardly, if at all, to have been affected by the experimental treatments.

Reasoning: Neither Abstract Reasoning (4) nor Deduction (12) was significantly affected by the active treatments (see Table 16). A similar result was observed for the Reasoning Composite, formed by summing the two reasoning measures (the same scale factor was assigned each component measure). Deduction did, however, tend to be impaired by the heavy alcohol dose (a result not easily attributed to an alcohol-induced reduction in visual acuity, for Ss had ample opportunity to vary their distance from the Deduction test protocol so as to compensate for impaired vision).

Drug effects on reasoning have been studied by other investigators. Thus, Flory and Gilbert (1943) found an analogies test to be unaffected by caffeine citrate. The relevance of Flory and Gilbert's findings for the present study is limited because they used a speeded reasoning test, and because of the absence of pre-drug values, which inflated experimental error in their study. Some impairment of reasoning has been reported for alcohol; see Mead (1939) and also see Jellinek and McFarland's discussion of the findings for arithmetic reasoning and for sorites, syllogisms, and conversions.

Reasoning appears to have been more resistant to drug effects than the functions discussed heretofore. Might this selective effect have been spurious, simply reflecting excessive experimental error

TABLE 16
TREATMENT EFFECTS ON REASONING

Measure	Placebo Mean	Difference Between Means of Active Treatment and Placebo		
		CAFFEINE	ALCOHOL (15.7 ML.-/M.^{2})	ALCOHOL (31.4 ML.-/M.^{2})
Abstract Reasoning (4)	1.29	0.29 4.06	1.14 4.06	0.45 4.23
Deduction (12)	1.21	−1.21 5.33	1.72 5.33	−4.50 5.71
Reasoning Composite	1.25	−0.46 3.47	1.43 3.47	−2.03 3.65

on the part of the reasoning measures? Not likely, for considerable testing time had been devoted to the reasoning tests, so that the Reasoning Composite might be about as precise a measure as, say, the Word Association or Associate Learning Composites.

In accounting for the reasoning measures' lesser susceptibility to drug effects, one might suppose that the rational processes shared in common by tests of inductive and deductive reasoning were simply little affected by the present doses. Furthermore, inductive and deductive processes may be sufficiently distinct that combining measures reflecting both processes might have served only to obscure drug effects on reasoning.

Another explanation emphasizes the fact that Abstract Reasoning and Deduction were "power tests." Most Ss finished these tests (Deduction in particular) before the time limits expired. Contrast this with the tests discussed in previous sections, which either had speeded time limits or else required Ss to handle a rapid flow of material. It may be that the present doses primarily affected Ss' capacity to work at a rapid pace, irrespective of the specific task performed, while leaving unaffected Ss' capacity to work at a leisurely pace.

Flexibility of Thinking: Several of the present study's tests called for flexibility of thinking (here regarded as being akin to the Integration III factor identified in World War II Air Force studies [Guilford, 1947]). These tests required thinking through solutions to novel problems, or else mental manipulation of ideas or other conscious contents pertaining to diverse aspects of a complex situation.[11]

Match Problems (6) called for a restructuring of visually presented situations. This restructuring tends to take place in the "mind's eye." It also tends to have an all-or-none character: that is, the sought-for situation "snaps into focus," or it doesn't; S

---

[11] Note that the "flexibility of thinking" discussed here, requiring *agility in the manipulation of ideas* or other conscious contents, differs from the "flexibility" involved in Digit Symbol (17), which calls for a harmonizing of different activities or functions (see the section: *Visual-motor co-ordination*, above), as well as from the "flexibility" involved in the Stroop Color-Word Test (18), which calls for conscious suppression of a deeply ingrained response tendency in favor of a more superficial tendency (see the section: *Resolution of competing tendencies*, Chapter 7).

TABLE 17

TREATMENT EFFECTS ON FLEXIBILITY OF THINKING

Measure	Placebo Mean	Difference Between Means of Active Treatment and Placebo		
		CAFFEINE	ALCOHOL (15.7 ML.-/M.^{2})	ALCOHOL (31.4 ML.-/M.^{2})
Digits Backward (3 trials) (3)	1.21	−0.57 2.30	−1.00 2.30	−1.64 2.35
Continuous Subtraction (19)	−3.86	−0.50 3.05	1.00 2.98	0.64 3.34
Arithmetic Errors (9)	2.00	0.07 2.91	−1.79 2.91	−2.43 2.91
Match Problems (6)	2.21	0.79 1.46	0.14 1.46	0.64 1.54
Visual Adaptive Flexibility Composite	1.36	1.96 2.94	0.18 2.94	0.07 3.09
Adaptive Flexibility Composite	1.57	1.33 2.08	−0.48 2.08	−0.76 2.08

"sees" it, or he doesn't. Match Problems was unaffected by the active treatments (see Table 17). This finding is inconclusive, as the measure available for present use was not highly reliable.

Gottschaldt Figures (22), which required another kind of restructuring of visual stimuli, has been considered in the section: *Visual thinking* (above). This measure was not significantly affected by drugs. Significant findings were likewise absent for the Visual Adaptive Flexibility Composite, formed by combining Gottschaldt Figures and Match Problems (for the scale factors assigned these component measures, see Table 3).

Arithmetic Errors (9), a more complex version of the Sign Changes tests devised by Guilford and associates (cf. Kettner *et al.*, 1959), called for the reformulation of mathematical identities by substitution of one elementary mathematical operation for another (see Appendix B). The task, which had some resemblance to the balancing of chemical equations, is a challenging one. The best Arithmetic Errors performances are those of Ss able to survey a problem's diverse aspects without becoming lost in the process. Although this intuitive approach appears to be that most commonly used by successful Ss, good performance is also possible using a logical approach, eliminating incorrect logical alternatives in rapid, systematic, trial-and-error fashion.

Though not quite significant, the heavy alcohol dose effect on Arithmetic Errors was surprisingly large, considering the brief, exploratory form of the test employed. A slightly more reliable form of test might have been affected significantly. Arithmetic Errors was added to the component measures of the Visual Adaptive Flexibility Composite to form the more general and more heterogeneous Adaptive Flexibility Composite (see Table 3); this more inclusive measure was not significantly affected by the active treatments.

Continuous Subtraction (19) and Digits Backward (3) also have a bearing on flexibility of thinking; while requiring no problem solving, these are challenging tasks, calling for mental manipulation of conscious contents. The non-significant Continuous Subtraction findings are better understood against the background of Goldberg's (1943) findings for a similar continuous subtraction measure. Studying the effects of alcohol in large quantities, Goldberg found continuous subtraction to be the only one of a number of measures whose impairment was not clearly dependent on blood alcohol concentration. The practice effect for Goldberg's continuous subtraction measure was so large that it almost obscured the drug effect; only for high blood alcohol concentrations was the alcohol-induced impairment of continuous subtraction clearly distinguishable from the practice effect. Goldberg's findings as well as informal observations made during the present study suggest that low test reliability accounted for the present lack of significant Continuous Subtraction findings.

Perhaps the most precise of the present study's flexibility of thinking measures was Digits Backward. Here, rote material was presented to Ss orally, at a predetermined rate. After apprehending the presented material, Ss were to hold the acquired information in mind, while reproducing the digits in an order opposite to that in which they were presented.

Comparison of Digits Backward findings with those for Digits Forward (2) is instructive, as the "backward" task is an elaboration of the simple retentive "forward" task (it is precisely this elaboration that enables Digits Backward to serve as an indicator of organic brain disturbance [Wechsler, 1944]). Although a combined analysis for Digits Backward and Forward failed to reveal a *significant* interaction between drugs and tests (see the section: Learning, above), the contrast between caffeine effects on these two measures is of interest: caffeine impaired Digits Backward while enhancing Digits Forward. Similar non-significant findings were obtained by Lehmann and Csank (1957). But Gilliland and Nelson (1939) found a (non-significant) improvement of digits backward with caffeine (these investigators did not study digits forward).

Caffeine's failure to enhance Digits Backward and the possible selective effects of caffeine on "forward" and "backward" rote memory tasks suggest some interesting speculations. Preceding sections have indicated that associative productivity is increased, and immediate, intentional recall improved, by caffeine. While each of these two caffeine effects may be considered an "enhancement of function," tasks calling for *both* association and recall *need not* benefit from these two functions' simultaneous enhancement. Thus, Ss' ability to hold in mind a sequence of digits while manipulating their order might have been interfered with by extraneous thoughts excited by caffeine. Such a disruption, due to a caffeine-induced facilitation of association, might have more than counteracted any beneficial caffeine effects on immediate, intentional recall. Limited support for this explanation is provided by Hull's (1935) finding of a significant increase in the number of anticipatory *intrusions,* while a series of nonsense syllables was being learned.

It is surprising to discover so few drug effects for these varied flexibility of thinking measures. The investigator had expected that the organism's energy reserves and maneuvering capacity would be augmented by caffeine and diminished by alcohol. Perhaps this expectation was not well founded. Or perhaps (except for Digits Backward, for which a special hypothesis is offered) the paucity of these findings is attributable to the relative unreliability of the measures available for use in this study. Note that the flexibility of thinking measures had time limits briefer than those

Intellectual Functioning

TABLE 18
Treatment Effects on Critical Judgment

Measure	Placebo Mean	Difference Between Means of Active Treatment and Placebo		
		CAFFEINE	ALCOHOL (15.7 ML.-/M.^{2})	ALCOHOL (31.4 ML.-/M.^{2})
Language Usage (26)	−0.21	0.57	2.71	−0.93
		4.64	4.64	5.38
Time Estimation: I (37)[a, b]	0.29	0.24	0.01	−0.06
Time Estimation: II (37)[a, b]	0.93	−0.22	0.01	0.02
$\log \left(\frac{\text{Slow Writing Estimation}}{11 \times \text{Slow Writing}} \right)$ (36)[a]	−0.19	0.01	0.21	−0.08
		0.45	0.45	0.54
Digit Symbol: Attainment Discrepancy (17)[a]	1.36	−0.71	−2.36	−0.11
		4.97	4.97	5.56
Digit Symbol: Predicted Ceiling $\left(\frac{\text{Self}}{\text{Average}} \right)$ (17)[a, b]	1.148	−0.002	−0.002	−0.037
Estimation of Treatment Effects (34)[a]	0.07	−0.14	−1.07	−2.00
		2.80	2.80	3.38
Estimation of Treatment Effects, Corrected for Actual Effects (34)[a]	−12.4	−25.9	−24.3	49.3
		63.0	63.0	76.0
Whiskey Estimate (35)[a, b]	7.43	2.61	13.64*	18.82*
Whiskey Estimate, Corrected (35)[a]	11.3	4.9	13.9 *	19.3 *
		12.8	12.8	15.4
Practical Judgment: Understanding How to Handle an Emergency (30)[a]	41.79	−0.86	0.29	−1.79
		9.94	9.94	11.12
Contingencies (24)	−0.43	−1.36	−0.64	−3.43*
		2.91	2.91	3.25

[a] Pre-drug values lacking.
[b] Differences between means evaluated by H-test.
* Significant at 0.05 level.

for any set of measures other than those for "visual thinking" (compare the section: *Test length,* Chapter 11).

Critical Judgment: Results for the very heterogeneous group of "critical judgment" measures (see Table 18) failed, in most instances, even to approach significance. An exceptional measure was Contingencies (24), which required Ss to anticipate needs arising in novel situations. Contingencies was markedly impaired by the heavy alcohol dose.[12]

The paucity of significant critical judgment findings is particularly surprising in view of Jellinek and McFarland's conclusion that "... the judgment of time is greatly impaired. ... subjects made wrong judgments of their own performances. The few investigators who maintained that self criticism was not injured by alcohol were those who carried out experiments on themselves and thus were making judgments of their own judgments. ... on the whole, the impairment of judgment through alcohol cannot be questioned but ... no estimate can be made of the degree of impairment. Experimentation on the effects of alcohol on different types of judgment ... is indicated." McFarland (1953) was even more explicit: "The first symptom [of alcohol intoxication] is a dulling of one's critical judgment. ... Finally, there is a loss of insight into the extent of one's impairment."

The dearth of critical judgment findings for the present study is probably largely accounted for by the following factors:

1. Experimental error was heightened because most critical judgment measures were brief or of an exploratory nature.

2. Experimental error was further inflated because post-drug values were corrected for the corresponding pre-drug values in only two instances (pre-drug values were lacking for the remaining critical judgment measures).

3. The illness of six Ss made it especially difficult to detect heavy alcohol dose effects on critical judgment. More than half the incapacitated Ss were eliminated from the experiment without having received a single test of critical judgment. Indeed, results for all six incapacitated Ss were missing for the many fixed-order critical judgment tests administered toward the end of session 4.

4. Experimental error for the many fixed-order critical judgment tests (administered after completion of session 4's test batteries) was probably smaller than for tests administered in bat-

---

[12] Compare the near-significant impairment of Arithmetic Errors (9) by the heavy alcohol dose. Like Contingencies, Arithmetic Errors usually involved anticipatory thinking: a problem situation was surveyed and Ss "came up with" a specified relationship.

teries, whose order of presentation was randomized. (Randomizing test order, while it eliminated one kind of bias, increased experimental error.) The end-of-session tests' slight advantage with respect to experimental error was undoubtedly more than offset by the fact that blood levels at the end of the session (when large quantities of drug had already been metabolized) were considerably reduced from their peak values. (Contingencies, definitely impaired by the heavy alcohol dose, was administered not at session's end, but as part of a battery whose order of presentation was randomized.)

The critical judgment measures' insensitivity to drug effects seems to have been more than a matter of excessive experimental error. For example, the heavily alcoholized Ss (and also the lightly alcoholized Ss) *correctly* provided higher estimates of the alcohol amount each had consumed during the experiment than did placebo and caffeine Ss (see the significant findings for Whiskey Estimate, Corrected (35),[13] Table 18; also see Appendix A). The heavily dosed Ss clearly retained *some* good judgment.

Ss' treatment effects estimates further suggest that critical judgment was not completely undermined and indeed may not have been greatly disturbed by the heavy alcohol dose. For though the 8 heavily dosed Ss remaining at the end of the experiment *underestimated* the extent to which their over-all performance had been impaired by alcohol (see Estimation of Treatment Effects, Corrected for Actual Effects [34],[13] Table 18), they did recognize that impairment had occurred (see Estimation of Treatment Effects [34], Table 18). These findings might well have been significant had they been obtained from all 14, and not simply 8 of the Ss who received the heavy alcohol dose. Recognition of impaired general functioning, after quantities of alcohol even larger than those administered in the present study, has been reported by Loomis and West (1958).

Relevant to the present discussion are the findings for Would Trust My Judgment (see Table 4 and Chapter 5). Here, as with

---

[13] Positive signs do not necessarily indicate enhanced judgment, nor negative signs impaired judgment, for subjective estimates of objectively measured quantities.

Estimation of Treatment Effects, Ss receiving the heavy alcohol dose recognized their impaired intellectual functioning. But, whereas they tended to *under*estimate treatment effects on overall intellectual functioning, the heavily alcoholized Ss seemed *overly ready* to concede that their judgment was impaired by alcohol; they were more pessimistic about their judgment than was warranted by the objective findings. They displayed poor judgment about their judgment, suffering from an impairment of "second-order judgment," as it were. This discrepancy between first- and second-order judgment effects (i.e., between objective and subjective findings for critical judgment) may have been due to the fact that Would Trust My Judgment self-ratings were obtained at higher blood alcohol concentrations (on the average) and for a larger sample of Ss than were the many objective critical judgment measures obtained toward the end of session 4.

The cumulated evidence suggests that critical judgment, while less affected by alcohol than say visual thinking, is impaired at least by large quantities of alcohol. This impairment is indicated less by present findings than by reports of other investigators. The resistance of the present critical judgment measures to impairment by alcohol seems attributable both to the design of the experiment and to the many missing values for critical judgment measures. Another factor is the tendency of highly intellectual Ss (such as those of the present study) to be put on their guard when questioned closely; impairment might more readily have been revealed by indices of spontaneous judgment than by answers to carefully formulated questions. Critical judgment may conceivably be more severely impaired by alcohol in Ss' native habitat (driving home from a tavern, for example) than in an artificial laboratory environment.

Discrimination of Significant Features in Practical Social Situations: Related to critical judgment is the capacity to discriminate significant features in practical social situations, reflected by the first three measures listed under this heading in Table 19, as well as by Practical Judgment: Understanding How to Handle an Emergency (30) (see the section: *Critical judgment,* above). These four measures, considered separately, or in combination as the Practical Judgment Composite (see Table 3), failed to reveal significant drug effects (see Table 19).

TABLE 19
Treatment Effects on Discrimination of Significant Features in Practical Social Situations

Measure	Placebo Mean	Difference Between Means of Active Treatment and Placebo		
		CAF-FEINE	ALCOHOL (15.7 ML.-/M.^{2})	ALCOHOL (31.4 ML.-/M.^{2})
Picture Completion (27)[a]	17.00	0.00 2.31	−0.21 2.31	0.50 2.58
Evaluation of Arguments (29)[a]	10.29	5.07 6.11	1.86 6.11	2.00 7.08
Practical Judgment: Understanding the Reasons for Accepted Social Practices (30)[a]	42.07	0.22 8.80	−1.14 8.80	−4.00 9.85
Practical Judgment Composite[a]	43.11	2.38 6.05	0.50 6.05	0.00 6.56

[a] Pre-drug values lacking.

How to account for this lack of response to drugs? Was discrimination of significant features in practical social situations simply unaffected by the present doses? Perhaps. But note that excessive sensitivity to drug effects had not been expected from the Practical Judgment Composite, whose component measures were not notably reliable, and moreover lacked pre-drug values. And recall that these measures were investigated under non-speeded conditions. The absence of significant findings for the Practical Judgment Composite tends indeed to support the hypothesis (see the section: *Reasoning*, above) that the present doses affected Ss' capacity to work at a rapid pace (irrespective of the specific task performed), but not Ss' capacity to work at an unhurried pace. This hypothesis gains further support from the dearth of findings for the critical judgment tests, which were largely administered under non-speeded conditions. Interestingly enough, it was Contingencies (24), possibly the only critical judgment test that many Ss failed to complete in the allotted time, that almost alone among the critical judgment tests showed significant drug effects.

7

AFFECTIVITY AND RESPONSE TO STRESS

AFFECTIVITY: Attention has thus far been concentrated on intellectual functioning. The focus of interest now turns from the cognitive sphere to the affective sphere. As with intellectual functioning, affective functioning was reflected by self-ratings, by capacity measures, and occasionally by expressive measures.

It will be recalled that several of the self-ratings—for example, Irritable; Impulsive, Lacking in Restraint; Interested in Taking the Tests—had a distinct bearing on affectivity. The doses administered in the present study did not appear to have any systematic effects on these self-ratings (see Table 4 and Chapter 5). The affective aspects of Ss' subjective state seemed even less regularly altered by the experimental treatments than its cognitive aspects.

Considered here is Sentence Completion: Expression of Feelings (28). The doses administered apparently had no effect on Ss' attitudes toward the open expression of their feelings (see Table 20). But note that this conclusion is based on a brief, exploratory measure, which lacked pre-drug values.

Reaction to Bodily Threat: Mere participation in the study may have proved stressful for some Ss. Measures were available, however, reflecting Ss' responses to two particular stress situations—the blood-taking procedure and the Stroop Color-Word Test.

Consider first Reaction to Puncture. Significant drug effects were revealed neither for Withdrawal nor for Tightening (these measures were transformed to logarithms; see Table 20) nor for a

Affectivity and Response to Stress 87

combination of the two, the Reaction to Puncture Composite (see Table 3 for the make-up of this composite). Nor were significant drug effects obtained for Dislike Puncture (see Table 4 and Chapter 5), although Ss' expressed antipathy to having their fingers punctured was perhaps more affected by drugs than the actual withdrawal or tightening of their fingers.

That both subjective and objective indices of reaction to bodily threat were essentially unmodified by drugs may seem surprising considering the analgesic action attributed to alcohol both by laymen and by scientific investigators. The evidence for an alcohol-induced increase in pain threshold is definite (Dodge and Benedict, 1915; Goldberg, 1943; Mullin and Luckhardt, 1934;

TABLE 20

Treatment Effects on Affectivity and Response to Stress

Measure	Placebo Mean	Difference Between Means of Active Treatment and Placebo		
		CAFFEINE	ALCOHOL (15.7 ML.-/M.^{2})	ALCOHOL (31.4 ML.-/M.^{2})
AFFECTIVITY:				
Sentence Completion: Expression of Feelings (28)[a, p] .	0.21	−0.29 1.51	−0.21 1.51	0.14 1.69
REACTION TO BODILY THREAT:				
log ([Reaction to Puncture: Withdrawal] + 2)[a, o]	0.69	0.05 0.44	0.02 0.44	0.01 0.44
log ([Reaction to Puncture: Tightening] + 2)[a, o]	0.87	−0.09 0.35	−0.16 0.35	−0.08 0.35
Reaction to Puncture Composite[a, o] .	0.78	−0.02 0.33	−0.07 0.33	−0.04 0.33
RESOLUTION OF COMPETING TENDENCIES:				
Stroop Conflict Time (18)	−8.2	−5.6 11.1	0.1 11.1	4.3 12.4
Stroop Ratio × 100 (18) . .	4.29	1.00 7.30	1.57 7.30	−5.79 8.17

(Continued on next page)

TABLE 20 (Continued)
TREATMENT EFFECTS ON AFFECTIVITY AND RESPONSE TO STRESS

Measure	Placebo Mean	CAFFEINE	ALCOHOL (15.7 ML.-/M.^{2})	ALCOHOL (31.4 ML.-/M.^{2})
REACTION TO STRESS (PROJECTIVE):				
Sentence Completion: Constructive Reaction to Stress (28)[a, p]	−2.36	2.21 3.87	2.79 3.87	0.43 4.33
Sentence Completion: Constructive Reaction to Separation (28)[a, p]	−1.71	0.43 2.35	0.79 2.35	−0.22 2.63
Rosenzweig Picture-Frustration: Striving to Overcome Frustration (32)[a, p]	4.25	−0.36 2.05	−0.14 2.05	−0.16 2.30
Constructive Reaction to Stress Composite[a, p]	0.91	0.79 3.19	1.36 3.19	−0.11 3.46
Sentence Completion: Mobilization of Aggressive Energies for Self-Forwarding Action (in Reaction to Stress) (28)[a, p]	−1.07	2.79 3.68	3.29 3.68	0.64 4.12
Rosenzweig Picture-Frustration: Expression of Aggression (32)[a, p]	19.36	−1.79 2.24	−0.36 2.24	−1.08 2.51
Rosenzweig Picture-Frustration: Extropunitiveness versus Intropunitiveness (32)[a, p]	8.21	−2.86 5.04	−0.29 5.04	−3.99 5.64
Consequences: Favorable Outcome (7)[p]	0.86	−0.64 2.77	−0.71 2.77	−0.57 2.77
Consequences: Unfavorable Outcome (7)[p]	−2.50	5.50* 4.59	3.93 4.59	1.00 4.59
Consequences: Active Efforts at Mastery (7)[p]	0.14	−0.50 2.39	0.07 2.39	−0.71 2.39
Consequences: Passivity (7)[p]	−1.57	4.79* 3.90	2.64 3.90	0.57 3.90
Consequences: Passivity, Corrected (7)[p]	−0.79	1.86 2.81	2.36 2.81	2.86* 2.81

[a] Pre-drug values lacking.
[e] Expressive measure.
[p] Projective measure.
* Significant at 0.05 level.

Wolff, Hardy, and Goodell, 1941), although some investigators have failed to observe alcohol's analgesic action (Carlson, Kleitman, Muehlberger, McLean, Gulliksen, and Carlson, 1934; Miles, 1924).

Drug effects on Dislike Puncture self-ratings and on Reaction to Puncture ratings were not, of course, a direct index of the drug's analgesic actions, since reaction to the blood-taking procedure involved much more than awareness of pain. Ss' responses were affected by their attitudes toward "mutilation." Finger tightening, which generally occurred before Ss were pierced by the blood lancet, was prompted by anticipation, *rather than* by actual awareness, of pain. Finger withdrawal on puncture, while dependent on awareness of pain, seemed to bear surprisingly little relation to the actual physical trauma. Many Ss were unpleasantly affected by the sight of their blood, and this attitude undoubtedly affected self-ratings for Dislike Puncture. The operation of these various factors might well have obscured alcohol's analgesic action.

Resolution of Competing Tendencies: Stroop Conflict Time and Stroop Ratio × 100 (18) were two non-independent measures of Ss' capacity to respond in a prescribed manner, despite distraction by a more deeply ingrained response tendency. Table 20 suggests that caffeine helped Ss integrate these competing tendencies, while the heavy alcohol dose reduced Ss' resistance to distraction; these findings lacked significance. (Note that a *plus* sign for Stroop Conflict Time has the same meaning as a *minus* sign for Stroop Ratio × 100.)

Related to these findings is an experiment by Hildebrandt (cited by Jellinek and McFarland). After the establishment of a set of paired associations, Ss were required to respond to the first member of each pair with a rhyme or with a reversal, in place of the already learned associate. Ability to break through the established associations was apparently unimpaired by an alcohol dose of 50 cc.; Ss were no longer able to maintain the desired set with a dose of 100 cc.

The Stroop findings as well as those of Hildebrandt indicate that small to moderate quantities of alcohol permit a response to be maintained by force of will, despite competition from more deeply ingrained responses. The same response can only be maintained at great cost, however, with quantities of alcohol appreciably in excess of those administered in the present study.

Reaction to Stress (Projective): Reaction to stress was studied not only by means of field observations (Reaction to Puncture), self-reports (Dislike Puncture), and objective performance tests (Stroop Color-Word), but also through the use of projective tests.

Constructiveness of response to stressful social situations was reflected by the first three measures listed under *Reaction to stress (projective)* in Table 20. (These exploratory measures, experimentally independent of each other, were intercorrelated positively; correlations between pairs of measures, averaged over all four treatments, ranged from +0.25 to +0.37.) Significant findings were lacking for these three measures, considered separately or in combination as the Constructive Reaction to Stress Composite (see Table 3). This lack of significant results is not conclusive, however, for these were relatively primitive and untried instruments, included in the study for exploratory purposes; moreover, experimental error was exaggerated due to the absence of pre-drug values.

The positive effect of the small alcohol dose and of caffeine on Sentence Completion: Constructive Reaction to Stress (28) is worthy of attention, despite the absence of significant findings. Similar in sign but more striking in magnitude were the results for Sentence Completion: Mobilization of Aggressive Energies for Self-Forwarding Action (in Reaction to Stress) (28). (Note that the two Sentence Completion: Reaction to Stress measures lacked experimental independence, though each was scored independently, as they were based on the same set of responses. The correlation between the two measures, averaged over all four treatments, was +0.76.) These results, which suggest that the small alcohol dose and also caffeine helped mobilize Ss for self-forwarding action in potentially disruptive situations, might well have been significant had pre-drug values been obtained.

Also related to the subject of aggression were two non-independent Rosenzweig Picture-Frustration measures—Expression of Aggression, and Extropunitiveness versus Intropunitiveness (32). Concerned with the arousal and disposition of *hostility* following frustration, these measures had different connotations from those of Sentence Completion: Mobilization of Aggressive Energies, which had to do with Ss' readiness to summon up energy reserves for self-forwarding action in situations threatening to block Ss' forward movement (Nash, 1958). The findings for the two Rosenzweig Picture-Frustration measures (see Table 20) suggest that hostile feelings may be diminished, or perhaps turned inward, by caffeine and by the heavy dose of alcohol; these findings were not significant.

Further information on response to difficult situations was provided by Consequences (7), whose standard measures (Low-quality and Remoteness) of intellectual tendency and/or capacity are considered in the sections: *Associative productivity* and *"Quality" of associations*, in Chapter 6. Consequences was also scored for "content" to provide information about the *kinds of thinking* indulged in by Ss faced with sudden deprivation. Responses were scored along two *non-standard* "dimensions": each response was scored as having either a Favorable or Unfavorable Outcome; also, as displaying either Active Efforts at Mastery, or Passivity. Scores for Favorable and Unfavorable Outcome were based on the same set of responses as those for Active Efforts at Mastery and Passivity. These two "dimensions" lacked experimental independence, and they were also found to be closely associated in practice; scores of Passivity and Unfavorable Outcome were quite frequently assigned to the same response, for example.

Table 20 indicates that drug effects on "opposing" scoring categories for the Consequences content dimensions were, with one exception, opposite in sign. Drug effects on "similar" scoring categories (Passivity and Unfavorable Outcome, for example) were, again with one exception, identical in sign and similar in magnitude. Findings for the Consequences content "dimensions" thus displayed a high degree of internal consistency.

Further examination of the Consequences results reveals marked drug effects for Unfavorable Outcome and for Passivity; scores for both scoring categories were increased by caffeine (this effect was significant) and by the mild alcohol dose (not significant). (Drug effects for Favorable Outcome and for Active Efforts at Mastery were negligible.) These findings are somewhat at variance with those for the study as a whole, for they suggest that caffeine had a depressing or immobilizing effect just when Ss most needed free access to their energy reserves. Could the Consequences content findings have been spurious, due perhaps to the lack of experimental independence among Consequences' three "dimensions" (Favorable–Unfavorable Outcome, Active Efforts at Mastery–Passivity, and Remoteness–Low-quality)? Might the drug effects for, say, Low-quality (see the section: *Associative productivity*, in Chapter 6) have been more genuine, or at least more fundamental, than those for, say, Passivity?

The existence of empirical intercorrelations between the three Consequences dimensions does not in itself demonstrate the causal primacy of any one of these dimensions. Passivity was, however, corrected by subtraction for Low-quality, on the assumption that Consequences' content findings were by-products of drug effects on associative productivity. The corrected Passivity findings (see Table 20) differed considerably from the uncorrected findings, in that Passivity, Corrected was much more markedly affected by the heavy alcohol dose than was the uncorrected score; the reverse was true for caffeine. Though more in accord with the results for the study as a whole, the corrected Passivity findings still strike a discordant note, implying as they do that some energies became less accessible to Ss following ingestion of caffeine or of the mild alcohol dose.

Summary of Results for Stress Reactions: Results for the various measures of reaction to stress were rather meager, in view of the considerable experimental effort concentrated on this area of functioning. Practical Judgment: Understanding How to Handle an Emergency (30) was unmodified by the active treatments (see the section: *Critical judgment,* Chapter 6). Reactions to *live* stress situations (Puncture and Stroop Color-Word) were hardly more affected by the active treatments, although Hildebrandt's findings indicate that the ability to break through established associations is impaired by alcohol doses exceeding those of the present study.

Associations evoked by *hypothetical* stress situations—Consequences, Sentence Completion: Reaction to Stress, and Rosenzweig Picture-Frustration—were more definitely affected by drugs, but the meaning of these findings is not entirely clear. Findings for related groups of these content or projective measures were surprisingly consistent, considering the exploratory nature and relative unreliability of these measures, and considering the absence of pre-drug values for all but the Consequences measures. Note the interrelations among Consequences measures as well as among Sentence Completion measures. And note the agreement between Sentence Completion and Rosenzweig Picture-Frustration measures.

Against this background of agreement there stands out the sharp disagreement between Consequences' content measures, on the one hand, and the Sentence Completion and Rosenzweig Picture-Frustration measures on the other. Take, for example, Sentence Completion: Mobilization of Aggressive Energies, and Consequences: Passivity—the findings for these theoretically *opposed* measures were actually similar rather than opposite in sign. Or consider the Constructive Reaction to Stress Composite, which was correlated positively—*not* with Consequences: Favorable Outcome, as might have been expected—but with Unfavorable Outcome.

Given the marked discrepancy between the Consequences content findings, on the one hand, and findings for Sentence Completion: Reaction to Stress and Rosenzweig Picture-Frustration on the other, what may we conclude about drug effects on hypothetical stress situations? It is by no means easy to decide between these discrepant sets of findings. The particular "hypothetical stress" measures employed in the present study were essentially untried and not too much can be said about their validity. The statistical significance of the Consequences findings might be taken as an argument for *their* primacy. But the agreement between Sentence Completion and Rosenzweig Picture-Frustration results suggests that *these* findings may be not at all capricious.[1]

---

[1] The Sentence Completion and Rosenzweig Picture-Frustration stress findings might simply have reflected pre-drug differences in the intellectual level of Ss assigned to the four treatments. This possibility is raised by the lack of pre-drug values for the projective measures in question, and by the somewhat superior Vocabulary scores of caffeine and small alcohol dose Ss (see the section: *Subjects*, Chapter 4). These projective measures were not correlated with Vocabulary score, however, suggesting that the Sentence Completion and Rosenzweig Picture-Frustration stress findings were not attributable to pre-drug differences between treatment groups.

The apparent contradictions among these stress measures are here resolved by appeal to the findings for intellectual capacity. It seems most prudent to ignore the Consequences content findings, which were in least harmony with the very clear-cut findings for over-all intellectual capacity. Greater confidence is inspired particularly by the Sentence Completion: Reaction to Stress findings, which suggest that Ss' energies were mobilized both by caffeine and by the small alcohol dose.

Insufficient attention has been paid to the effects of alcohol and caffeine on reaction to stress. This area of knowledge is clearly in need of further study.

8

REGRESSION ON BLOOD LEVELS AND ON CUSTOMARY CONSUMPTION OF DRUGS

ANALYSIS OF THE psychological results simply in terms of the experimental treatments themselves might conceivably have failed to exhaust the information available from the present study. After all, drug effects are more nearly related to drug concentrations at the sites of drug action than to drug dosages administered to Ss.

Alcohol determinations based on samples of finger-tip blood might have differed considerably from alcohol concentrations at the presumed sites of drug action, especially when alcohol absorption had not yet been completed. Testing had, however, been delayed for an appreciable time following ingestion to permit drug concentrations of the various body fluids to reach equilibrium. The blood alcohol determinations of the present study were then probably in close correspondence with the alcohol concentrations at the presumed sites of drug action.

The blood level analysis did supplement the treatment-centered analysis in some instances. Thus, with mildly and heavily dosed Ss combined into a single analysis, performance for the end-of-session trial of Writing Speed (1) was seen to deteriorate with increasing blood alcohol concentration; see Fig. 4.[1] A regression line fitted to the data represented in Fig. 4 had a slope of −0.317. The standard deviation of the slope was 0.110; t was 2.88 (df = 22). The line of regression on blood alcohol concentration had a slope clearly different from zero (p = 0.01). Since the blood level analysis made use of information on differences *within* dos-

---
[1] See footnote 4, Chapter 4.

age levels as well as on differences *between* dosage levels, it was possible to observe a more clearly defined regression on blood alcohol concentration than on alcohol dosage level.

However, the blood level analyses *more typically* failed to provide information not already revealed by analysis of variance of the experimental treatments. Consider the beginning-of-session trial for Writing Expansiveness (1) (see below; also see Fig. 5). The regression on blood alcohol concentration of Writing Expansiveness (beginning-of-session trial), while of the same order of significance as that for Writing Speed (end-of-session trial), was almost entirely attributable to the difference between means of the two alcohol dosage levels. The blood alcohol concentration regression analyses generally supported but infrequently supplemented the information obtained from analysis of variance of experimental treatments.

Also considered was the possible modification of psychological findings by repeated drug use. Laymen and scientific investigators alike believe that the acute effects of alcohol (Goodman and Gil-

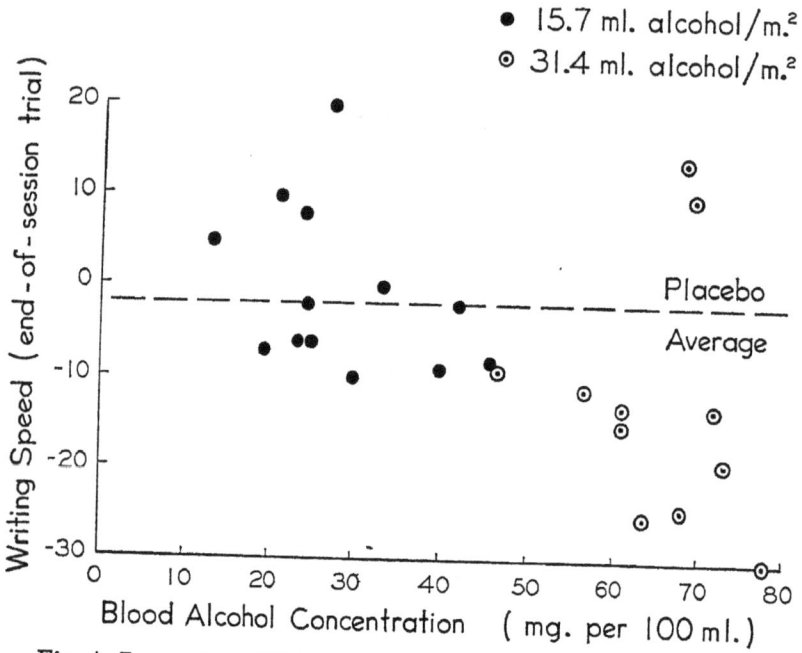

Fig. 4. Regression of Writing Speed on blood alcohol concentration.

man, 1955; Mirsky, Piker, Rosenbaum, and Lederer, 1941; Newman, 1941), and possibly those of caffeine (contrast the opinions of Landis [1958] and Goodman and Gilman [1955]), are reduced by habitual drug consumption. Significant regression of drug effects on estimated rate of customary drug consumption was, however, rarely revealed in the present study, either for alcohol or for caffeine.

The present regression findings are placed in clearer perspective by reference to Goldberg's (1943) findings. Goldberg administered rather large quantities of alcohol to fasting Ss. Three groups of Ss, differing in their drinking habits, were studied: abstainers and near-abstainers, moderate drinkers, and very heavy drinkers. Doses varied from 120 cc. absolute, the minimum dose administered to abstainers, to 340 cc., the maximum dose administered to any S. For the heavy drinkers at least, peak blood alcohol concentrations exceeded 100 mg. per cent, and at times, even 150 mg. per cent.

Goldberg made repeated measurements of various behavioral indices during the course of a single session. Values obtained for the blood alcohol curve's "ascending" part were treated separately from "descending" values. It is Goldberg's "descending" rather than his "ascending" values that are considered here, for the former were more numerous, provided more stable results, and are more comparable to the present alcohol results (generally obtained on the curve's "flat" or "descending" parts).

After transformation to logarithms, Goldberg's measures mostly showed a significant linear regression on blood alcohol concentration. Separate analyses were conducted by Goldberg for abstainers, moderate drinkers, and heavy drinkers. Regression lines for these three groups of Ss tended to be similar in slope but displaced from each other; x-intercepts (of the order of 50 mg. per cent for the moderate drinkers) regularly varied with drinking habits. The larger the quantity of alcohol customarily consumed, the higher the blood alcohol concentration at which drug effects were noticeable. Had Goldberg's regression analyses been conducted without regard to his Ss' drinking habits, regression on blood alcohol concentration would have been significant much less frequently.

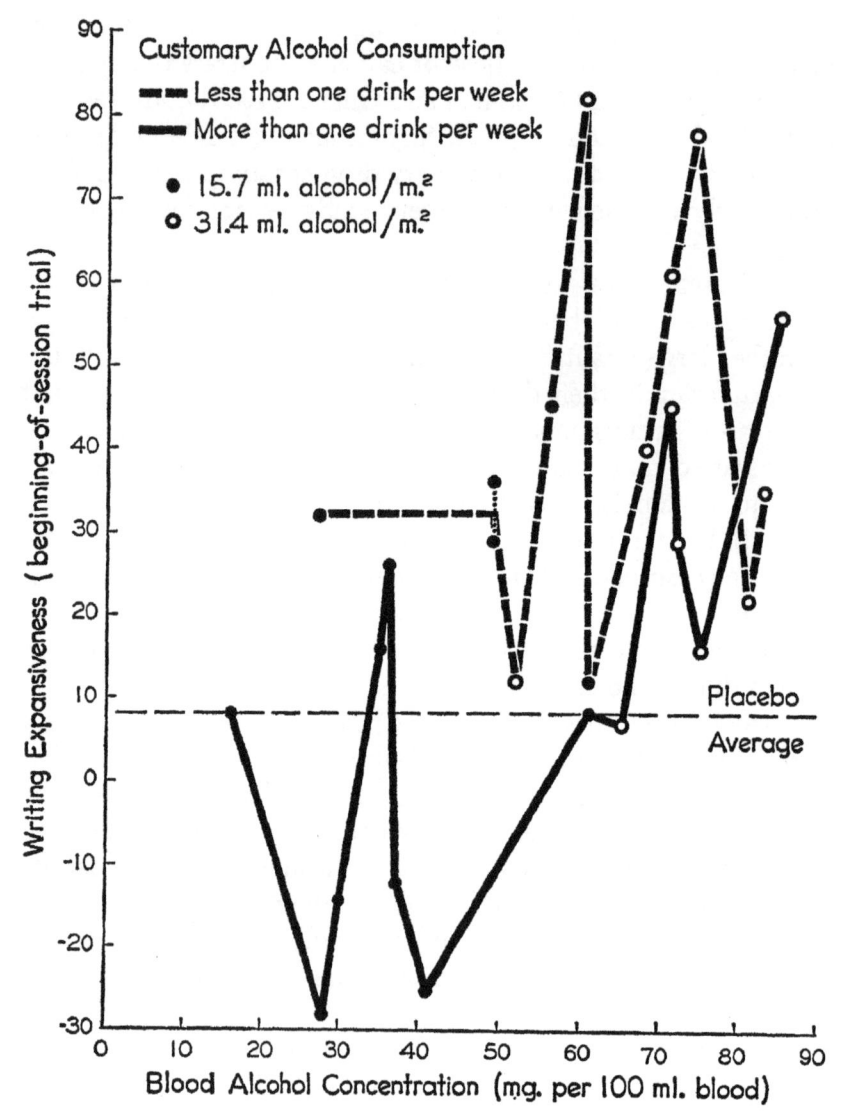

Fig. 5. Regression of Writing Expansiveness on blood alcohol concentration, as modified by habituation.

It is evident from Goldberg's study that the present regression findings were obscured because of the restricted range both of blood alcohol concentration and of Ss' drinking habits (compare the related observations of Vogel [1958] and of Drew, Colquhoun, and Long [1958]). Thus, with variations in drug effects most readily apparent at blood alcohol concentrations greatly in excess of 50 mg. per cent, regression on blood alcohol concentration would likely have been clearer had heavier alcohol doses been administered in the present study. (Note that missing values were most numerous for that treatment [the heavy alcohol dose] which almost exclusively gave rise to blood alcohol concentrations exceeding 50 mg. per cent.) And regression on customary alcohol consumption would likely have been better defined had the subject sample included heavier drinkers. At any rate, it should be remembered that the paucity of information added by the present regression analyses was a function of the design of the experiment, whose *primary* independent variables were neither blood levels nor drinking habits, but rather experimental treatments.

Present drug effects were, however, occasionally found to vary both with blood alcohol concentration and with customary alcohol consumption, despite the restricted range of these two measures. To demonstrate this multiple regression situation: Ss assigned to alcohol treatments were classified as "moderate" or "mild" drinkers, according to whether they customarily consumed more than one or less than one alcoholic beverage per week. The scattergram of Writing Expansiveness (beginning-of-session trial) versus blood alcohol concentration (see Fig. 5) was then entered, adjacent points being connected first for the moderate drinkers, then for the mild drinkers.[2] The curves for these two classes of drinkers intersected but once; it is evident that involuntary motor control was most disturbed among Ss least accustomed to alcohol. That regression on blood alcohol concentration was modified by tolerance to alcohol was most clearly evident for Writing Expansiveness (beginning-of-session trial), as this measure (obtained at near-peak blood levels) was among those most sensitive to alcohol.

---

[2] See footnotes 2 and 4, Chapter 4.

PART III
General Comments

9

FURTHER DISCUSSION OF PRESENT FINDINGS

IN PART II, the findings of the present and previous investigations were discussed function by function. A less compartmentalized approach is adopted in Part III, which explores certain more general implications of the findings.

Considerations Affecting Evaluation of Over-all Drug Effects: The following considerations affect evaluation of alcohol's and caffeine's over-all effects:

a) *Two-tailed Tests:* Previous findings that caffeine effects are largely beneficial and alcohol effects largely detrimental justify the use of one-tailed Dunnett tests for analysis of the present caffeine and heavy alcohol dose data. Two-tailed tests were applied to the small alcohol dose data, however, to permit detection of enhanced as well as of impaired functioning. For uniformity's sake, two-tailed tests were then applied to findings for *each* of the active treatments.

Caffeine and heavy alcohol dose effects have consequently been measured by an *overly conservative* yardstick. That these effects were actually more significant than is indicated by the tabulated results gives even greater weight to the already well documented conclusion that caffeine was largely beneficial, and the heavy alcohol dose largely detrimental, to over-all psychological functioning. Greater importance might well be attached to the borderline significant findings for caffeine and for the heavy alcohol dose but not for the small alcohol dose.

b) *Confounding Due to "Side Effects" of the Heavy Alcohol*

Dose: Caffeine effects on psychological functioning were fairly well defined. These effects were presumably mediated directly by the central nervous system (more than one form of direct c.n.s. mediation is possible, however—compare the section: *Barmack's hypothesis,* below).

Psychological findings for the heavy alcohol dose were about equally well defined, but their meaning (particularly their implications for c.n.s. functioning) was rendered ambiguous by "side effects," which ranged from a slight giddiness and a mild feeling of gastrointestinal disturbance to nausea and emesis. Although six Ss were excused from further testing when "side effects" had *clearly* disturbed their test-taking capacity, "side effects" were undoubtedly present in *milder* degree before testing of these Ss was interrupted. Since the final results include data obtained before these six Ss showed themselves to be clearly ill, the heavy alcohol dose findings reflect alcohol's "side effects" as well as its direct c.n.s. actions. Alcohol's "side effects" are undoubtedly also reflected by findings for the heavily dosed Ss who displayed no sign of illness and were able to complete the experiment.

Neither the examiner nor Ss receiving the heavy alcohol dose need have been clearly aware of "side effects" that did not obviously impair test-taking capacity. These milder "side effects" might nonetheless have exerted a subtle influence on the pattern of test results. Thus, mild subjective distress resulting from *direct* irritation of the gastric mucosa[1] might have proved distracting to Ss. If marked, such distraction could have prevented realization of boundary conditions required to preserve the univocal relation between test responses and the functions purportedly represented by these responses (Nash, 1959). If mild, such distraction could have confounded drug effects mediated directly by the central nervous system with those mediated by the gastric mucosa. Thus, though heavy alcohol dose effects were of the same order of magnitude as caffeine effects, the heavy alcohol dose

---

[1] The heavy alcohol dose, with a fluid volume twice that of the small alcohol dose, had correspondingly greater direct gastric irritant effects. This gastric irritation may, however, have been counteracted by peppermint oil, small doses of which apparently relax the pyloric sphincter (Sapoznik, Arens, Meyer, and Necheles, 1935) and relieve flatulence, nausea, and gastralgia (Osol and Farrar, 1950).

Further Discussion of Present Findings 105

findings have less clear implications for c.n.s. functioning than do the caffeine findings.

c) *Suggestion:* Drug findings may be biased or obscured by suggestion. This hardly seems to have occurred for the caffeine and small alcohol dose findings of the present study. Although the heavy alcohol dose findings could conceivably have been exaggerated because of suggestion, the impairment of function associated with the heavy alcohol dose appears to have been genuine —a direct consequence of the drug's own action—and to no great extent the product of suggestion.

Appendix A describes the double-blind procedures used in the present study, explains why procedures such as these were adopted, and evaluates their effectiveness in minimizing suggestion.

d) *Practice Effects and Confounding:* Sample problems were usually presented to Ss before capacity tests were administered, in order to minimize practice effects during the course of the experiment. That such practice effects did indeed occur is suggested by the tendency for capacity scores to improve after administration of placebo (see the placebo means[2] in Tables 7, 9, 11, 13, and 15 through 19).

In order to avoid confounding drug effects with the effects of factors such as practice or fatigue, observations obtained under the influence of the active treatments were compared with observations obtained under placebo control conditions. Since practice effects might often have continued during post-drug testing, drug effects on given functions might occasionally have been confounded with drug-induced changes in learning capacity.

The quantitative effects of caffeine and of the heavy alcohol dose on over-all intellectual functioning were clear-cut. Is the *meaning* of these findings as unambiguous? Might the over-all capacity effects of these doses have been artifacts, simply reflecting a drug-induced modification of practice effects, or do these findings provide information about drug effects on intellectual functioning in general?

It appears most unlikely that drug effects on over-all capacity

---

[2] Note that differences between post-drug and pre-drug values for placebo were a function not only of practice but also of imperfectly equated post-drug and pre-drug test forms.

merely reflected drug effects on learning capacity, even had marked drug-induced changes in learning occurred. For the over-all findings combined the results of many measures, representing a variety of functions; confounding would likely have been a problem only for those particular measures displaying marked practice effects. Furthermore, drug effects on learning, rather than having been marked, actually appear to have been milder than those for several other functions. Caffeine and heavy alcohol dose effects on over-all functioning therefore appear to have been genuine, and not simply artifacts reflecting the influence on drugs on learning.

Barmack's Hypothesis: A most interesting hypothesis regarding the psychological mode of action of c.n.s. stimulants such as caffeine, amphetamine, and ephedrine has been proposed by Barmack (1938, 1940). Barmack observed that performance on a continuous, repetitive adding task (see the section: *Fixed associations, overlearned,* Chapter 6) was *initially* unaffected by caffeine. Adding rate increased with the passing of time, presumably due to practice, but then declined as Ss grew weary and lost interest in the task. Impairment of addition as well as development of fatigue and boredom were retarded, if not entirely forestalled, by caffeine.

These experimental findings suggested to Barmack that the effects of the stimulants in question are "not particularly on the ability but on the inclination to do continuous repetitive work." Boredom involves, in Barmack's view, a reversion to the sleep state; caffeine, whose anti-hypnotic and alertness-inducing properties were emphasized by Barmack, acts to allay the development of unfavorable attitudes, such as boredom, toward the adding task. Caffeine effects on intellectual functioning are, according to Barmack, more likely attributable to their supporting influence on the alertness adjustments to intellectually challenging tasks, than to any direct drug effects on specific neural processes immediately involved in the exercise of intellectual capacities such as addition.

The drug effects on adding rate reported by Barmack do appear to have been a by-product of drug effects on boredom and fatigue. But Barmack's results, which hold for the specific conditions obtaining in his experiments, do not warrant the further conclusion that a specific *capacity*, as for example adding capacity, is *generally unaffected* by caffeine.

There is some doubt whether Barmack's adding task, though presented to his Ss as a test of capacity, actually met the requirements for such a test. Barmack's adding task understandably failed to arouse Ss' enthusiasm and to engage their full attention. This is evident from *placebo* Ss' reports that *10 to 20 per cent* of the time allotted for adding was spent daydreaming! Even during the first 15 minutes of the two-hour adding session, when placebo Ss were still fresh, they reported spending more than 10 per cent of their time daydreaming! (Many of Barmack's Ss had experienced the monotonous nature of the task before receiving placebo, for Barmack employed balanced or approximately balanced experimental designs. These Ss therefore had an opportunity to anticipate the task's monotony even at the beginning of their placebo session.)

Contrast the attitude of Ss in the present study, all of whom approached the five-minute addition task as if primed for a short-distance run; the present task was begun without delay and completed without daydreaming or other distraction.

The boredom experienced by Barmack's Ss clearly disturbed the boundary conditions required to preserve a univocal relation between adding responses and the capacity purportedly represented by these responses (Nash, 1959).

That performance is indeed enhanced by caffeine in the absence of boredom or other conditions unfavorable to the maintenance of alertness is suggested by the results obtained with the present brief adding task. Central nervous system stimulants such as caffeine are able not only to allay deterioration of performance but also to enhance performance, as by direct stimulation of specific capacities or by sharpening the focus of Ss' attention.

The extent to which intellectual effects of c.n.s. stimulants are mediated by drug effects on mood, on attention level, and on specific capacities depends, of course, on the particular conditions under which Ss are examined. The question raised by Barmack's work, whether the intellectual effects of caffeine are more critically dependent on some psychological factors than on others, is a worthy one and merits further exploration.

Practical Versus Statistical Significance: Drug effects have here been evaluated in terms of their statistical significance. But statistical significance depends on factors such as sample size and is not in itself a criterion providing ultimate answers to the often practical questions that stimulate psychopharmacological investigations. How, then, are statistical evaluations to be translated into practical terms? For example, does caffeine's statistically significant enhancement of various types of performance make an *actual* difference in everyday life?

Such a question has no simple answer. The practical implications of a set of statistical results depend on the particular objectives in the questioner's mind. Thus, the investigator might ask specific questions such as these:

1. Should an individual undertake responsible work requiring critical judgment after, say, three or four highballs, taken on an empty stomach? That such a dose would worsen the judgment of an abstainer, or even of an occasional drinker, is clear. The present findings do suggest, however, that critical judgment may be relatively resistant to impairment (compared, say, to visual thinking) among intelligent Ss aware of alcohol's potentially harmful effects on critical capacity.

2. How saturated with alcohol must an automobile driver (an occasional drinker, for example) be before he becomes a public menace? Visual function and visual-motor function were found to be (statistically) significantly impaired at blood alcohol levels short of 100 mg. per cent. Whether such impairment by itself constitutes a public menace cannot be determined from the present findings alone.

Much competent research has been directed specifically to the question of driving under the influence of alcohol. Applied by agencies charged with regulating highway traffic, this research has led to widespread adoption of quasi-legal standards of intoxication, the most common standard being 150 mg. per cent. But susceptibility to impairment by alcohol is not simply a function of blood alcohol concentration; and it varies markedly from individual to individual (Newman, 1955). Establishment of a general standard of alcoholic intoxication, while of practical value to law enforcement officials, requires decisions that can only be viewed by experimental psychopharmacologists as arbitrary and open to controversy.

3. Do small amounts of alcohol promote creativity? This is a controversial question. Many authorities state categorically that alcohol has *no* beneficial effects on the intellect. But the present findings suggest that alcohol can induce a freer flow of ideas. That these ideas are of uniformly poor quality seems doubtful. While creative solutions to difficult problems are unlikely to be conceived and fully elaborated under the influence of large quantities of alcohol, more moderate quantities of alcohol may shake one's everyday, unquestioned views, or otherwise render permeable the boundaries of previously fixed belief. Such altered ideas, even if not acted on at the time of intoxication, may provide a basis for later constructive action.[3] But creative effects such as these may not readily lend themselves to statistical demonstration.

4. As for caffeine, it appears to facilitate addition, either in short concentrated spurts or over prolonged periods of time. Might it not then be sound financial practice to encourage one's bookkeeper to ingest caffeine in moderate quantities? (Whether coffee would also be adequate is a separate question.) Might a mild dose of alcohol serve almost as well as caffeine? There are grounds for answering these questions in the affirmative; but final answers would depend on a host of considerations, such as permissible side effects, placebo effects, effects of doses exceeding the optimum, and risks of addiction.

---

[3] Compare Masserman's (1946) observations on alcohol and experimental neurosis. "In most animals . . . neurotic behavior returns in great part when the pharmacologic effects of the alcohol wear off. A few animals, however, seem able to utilize their temporary bravado during mild intoxications to 'work through' [their] conflict and so gradually re-establish more nearly 'normal' response patterns."

10

ALCOHOL LEVELS AND DIRECTION OF DRUG EFFECTS

THE PRESENT experiment was designed, in part, to provide information about alcohol effects as a function of alcohol level. Alcohol dosage level was indeed one of the experiment's independent variables. Blood alcohol levels were also determined; intended as a secondary feature of the experiment, the blood alcohol determinations added little to the information provided by the dose-response analyses.

The findings of previous analyses of regression on alcohol level are here considered together with the present findings, and the implications of the combined findings are examined.

Drug Effects as a Function of Alcohol Level: *Linear* regression of Miles Motor Driving Trainer performance on blood alcohol concentration was reported by Drew, Colquhoun, and Long (1958), who observed that tracking error and steering wheel movement each varied directly with blood alcohol concentration; the observed regression was significant, at the 0.001 level in each instance.

Non-linear regression on blood alcohol concentration has, however, been reported by other observers. Graf and Straub (cited by Goldberg) were each able to fit a second-degree curve to their data. Goldberg, who observed a number of measures under varying degrees of alcohol tolerance (see above, especially Chapter 8), also found that alcohol-induced impairment increased at an ever growing rate with blood alcohol concentration. But Goldberg was convinced that a significant linear regression on blood alcohol concentration was more nearly attained by a logarithmic than by a square-root transformation of the data. Logarithmic transformation appeared satisfactory for fitting points for the rising as well as for the declining portions of the blood alcohol curve (Goldberg fitted separate regression lines for the two portions of the curve). The form of regression observed for flicker fusion by Granit and Harper, by Crozier and associates, and by McFarland and Halperin (cited by Goldberg) appeared to be in accord with that observed by Goldberg.

The linear findings can probably best be regarded as a special case of the non-linear findings. Thus, Goldberg's non-linear findings were obtained with blood alcohol concentrations at times twice the magnitude of those associated with the linear findings of Drew *et al.* May not a straight line be fitted, approximately, to the initial portion of a second-degree or exponential curve?

That alcohol effects also increase non-linearly with *dosage level* is obvious from the present finding that over-all effects were negligible for the small alcohol dose but marked for the heavy alcohol dose. Drew *et al.*, who analyzed regression on blood level, also presented dose-response data indicating a non-linear regression of steering wheel movement on alcohol dosage level. Vernon *et al.* (1919) observed a linear regression on dosage levels ranging from 30 to 60 cc. absolute alcohol. But the fitted line, when extrapolated, had an x-intercept of about 27 cc., which led Vernon *et al.* to conclude "that the curve is not strictly linear when it approaches the abscissa, but is curved." Regression on alcohol dosage level seems to be described either by a second-degree curve or by an exponential curve.

The data of Drew *et al.* also indicated a *sigmoidial* increase in tracking error with dosage level: the dose-response curve first rose at an ever increasing rate (as with a second-degree or exponential curve), then reversed its course, flattening out to form a plateau. This finding suggests that a limit of impairment may exist for some measures (as when Ss become utterly incapable of responding) which is invariant in the face of further alcoholization. It is also suggested that second-degree and exponential curves, themselves more general than a straight line, may simply be segments of a sigmoidal curve.

Biphasic Alcohol Effects and the Disinhibition Hypothesis:
Dose-response curves and blood-alcohol-response curves need not be restricted to monotonic functions, as are the curves discussed above. Early investigators of the effects of small alcohol quantities reported that the dose-response curve reversed its direction after reaching a peak of enhancement, as dosage continued to increase, until the initial gains were cancelled; impairment of function was observed with still further alcoholization. Whether such *biphasic* alcohol curves do indeed exist has been the subject of some controversy.

"Enhancement" by alcohol has been reported for several functions. A shortened latent time of the patellar tendon reflex, observed by Travis and Dorsey (1929), appears to have been due to retardation of higher c.n.s. centers, with consequent release from inhibition of lower centers.

Findings of a shortened simple reaction time, immediately following ingestion of alcohol, have been reported. Jellinek and McFarland, though attributing *some* of these findings to inadequate data analysis or to confounding of drug effects with practice effects, did not exclude the possibility of an initial reduction of reaction time with alcohol doses of the order of 10 cc. absolute.

Occasional enhancement of function by alcohol doses up to 40 cc. absolute has been observed with devices requiring prolonged muscular effort—the ergograph, for example. Alcohol is much more likely to *mitigate fatigue-induced impairment*, however, than actually to *improve* performance on a prolonged muscular task. A critical review of proposed explanations for alcohol effects on prolonged muscular work has been provided by Jellinek and McFarland.

Small alcohol quantities have been observed to enhance not only simple motor and sensori-motor tasks but also rate of association. The measures entering into the Associative Productivity Composite were rather consistently (yet not significantly) enhanced by the small alcohol dose. This dose produced a near-significant enhancement both of Addition, Correct (8) and of Sentence Completion: Word Count (28).

Perhaps the most obvious explanation for the enhancement of association rate by the small alcohol dose (and for the biphasic

Alcohol Levels and Direction of Drug Effects 113

association curve suggested by these and other findings; see the section: *Associative productivity*, Chapter 6) is that alcohol in small quantities acts as a c.n.s. stimulant with respect to association. Any such suggestions have traditionally been opposed by the weight of scientific opinion. As recently as 1950, the U. S. Dispensatory maintained: "It is no longer believed that alcohol is a cerebral stimulant; all evidence indicates, to the contrary, that it is a depressant, diminishing intellectual activity" (Osol and Farrar, 1950). This view has been expressed even more forcefully by Goodman and Gilman (1955): "It may be stated categorically that alcohol is not a stimulant, but rather a primary and continuous depressant of the nervous system. ... all the results [for mental and physical abilities] indicate decreased efficiency."

The unequivocal position represented by these statements is based partly on the assumption, here designated the "disinhibition hypothesis," that instances of "apparent stimulation [result] from the unrestrained activity of lower [nervous] centers freed by the depression of higher inhibitory control mechanisms" (Goodman and Gilman). Accounting for impairment as well as for "apparent" enhancement of performance by alcohol, "disinhibition" is widely regarded as being of more general explanatory value than "stimulation." Rooted in the hierarchical view of c.n.s. functioning associated with the name of Hughlings Jackson, the disinhibition hypothesis gains much of its force from observations on the stages of general anesthesia, in which a central nervous system paralysis descends along the cerebrospinal axis, *except* that the medulla is temporarily bypassed in favor of the spinal cord, and lower spinal centers may be affected before upper spinal centers.

The applicability of the disinhibition hypothesis to small alcohol dose effects on association rate is suggested by several lines of evidence. The quickening of association with alcohol brings to mind the facilitation of speech, by sodium amytal for example, in narcotherapy. But narcotherapeutic doses produce central depression short of basal anesthesia, while the *average* blood levels induced by the small alcohol dose (clearly below 50 mg. per cent) are associated with a mild subclinical intoxication. Such a mild intoxication, rather than disturbing the relation between centers at distinctly different levels along the cerebrospinal axis, might perhaps produce its selective effects merely by altering the balance among *co-ordinate* centers at the highest levels of c.n.s. functioning. Although the disinhibition hypothesis could yet be salvaged by omitting its hierarchical features, such a hypothesis-saving move would clearly weaken the implications of the hypothesis.

The view that alcohol releases lower c.n.s. centers from the critical control ordinarily exercised by higher centers implies that quality of association is degraded by alcohol. This implication appears to be supported by reports of more superficial association (greater frequency of "clang" responses [see Jellinek and McFarland]; but the statistical significance of these findings is uncertain) and by the Word Count findings (which suggest that, under the influence of the small alcohol dose, Ss got lost in words and were unable to get to the point).

The disinhibition hypothesis further appears to suggest that self-criticism, which might ordinarily limit addition rate, is reduced by alcohol. With caution thrown to the winds, as it were, Ss might add more rapidly, while paying less attention to the accuracy of their work and neglecting to check their results. An increase in addition rate such as was observed with the small alcohol dose could only be purchased, according to the disinhibition hypothesis, at the expense of a deterioration in quality of performance.

But the small alcohol dose's Addition findings showed improvement in *quality* as well as in quantity. While the qualitative findings were not significant, they at least clearly refute the idea that quantitative enhancement of association by alcohol must be accompanied by degradation of quality.

Although future investigators might fail to confirm the present Addition findings, these findings, plus the fact that the small alcohol dose's association findings closely resembled those for caffeine, raise a serious question regarding the unlimited applicability of the disinhibition hypothesis.

11

TASK FEATURES AFFECTING RESPONSE TO DRUGS

A NUMBER of psychological measures were enhanced by caffeine, a similar number impaired by the heavy alcohol dose. Since all measures were not equally affected by a given dose, each dose appears to have exerted *selective* as well as *general* effects.

The question arises whether drug effects can be ordered along some sort of "susceptibility gradient" (Payne, Moore, and Bethurum, 1952). As has been seen, the stages of general anesthesia (and, more specifically, the disinhibition hypothesis) suggest that a gradient of susceptibility to general anesthetics exists along the cerebrospinal axis. In considering the *entire range of actions* produced by a general anesthetic, the law of descending inhibition provides a good first working hypothesis, despite the fact that behavior associated with the medulla and with the spinal cord does not conform to this hypothesis. The law of descending inhibition is less pertinent, however, *when comparing behaviors each associated with the highest levels of c.n.s. functioning*.

An understanding of drug actions can also be sought by considering which task features are favorable, and which unfavorable, to the manifestation of drug action. Are drug effects revealed, or perhaps induced, by certain kinds of tasks, and not by others?

Task Complexity: Psychological deficit following various kinds of acute impairment of c.n.s. functioning has been related to task complexity. Thus, anoxia was found by McFarland (1939) to lead to impairment first, of such functions as critical thinking and judgment; then, of speed and accuracy of computation; and finally, of sensori-motor processes. Most sensitive to impairment by anoxia, according to Grether (cited by Takala, Siro, and Toivainen, 1958), were mental arithmetic and code substitution; next most sensitive were tasks involving dexterity; least sensitive were Letter Maze and sensory discrimination. These findings provide broad confirmation for the view that "cognitive" impairment (by anoxia, at least) is a function of task complexity.

Findings for c.n.s. depressants other than alcohol are less consistent with the task complexity hypothesis. Steinberg (1954) found that impairment of psychological functions by nitrous oxide was accounted for by task complexity, but only when motor functions were neglected. Impairment of function observed by Goodnow, Beecher, Brazier, Mosteller, and Tagiuri (1951) and by Von Felsinger, Lasagna, and Beecher (1953) for the barbiturate pentobarbital sodium was *not* accounted for by task complexity.

The idea that mental processes are affected by alcohol according to their differential complexity was early espoused by McDougall (cited by Steinberg [1954]). Jellinek and McFarland, reviewing the earlier experimental literature for alcohol, observed: "On the whole, the tracing of pathways and the marking tests show considerably greater impairment than the less complex tests of eye and finger movements.... The greater the muscular coordination involved in a task, the greater is its impairment through medium dosages of alcohol." Noting similar trends for recall and for learned associations, these reviewers concluded that the hypothesis of task complexity suited the alcohol findings rather well, provided the sub-hypothesis of task familiarity (see below) was formulated to account for exceptions.

The present heavy alcohol dose findings are not satisfactorily accounted for in terms of the task complexity hypothesis. The heavy alcohol dose impaired the Visual Thinking, Digit Symbol (17), and Immediate, Intentional Recall Composites, as well as Contingencies (24). Stroop Ratio × 100 (18), Deduction (12), and Arithmetic Errors (9) also tended to be impaired. These measures were based on rather complex tasks. Other measures impaired by the heavy alcohol dose, such as the Perceptual Speed Composite (21) and Writing Expansiveness (1), were based on simpler tasks. Digits Backward (3), while more complex, was hardly more impaired than Digits Forward (2). The Practical Judgment Composite and Abstract Reasoning (4), both indices of complex processes, did not even show a tendency to impairment by the heavy alcohol dose. The relative ineffectiveness of the heavy alcohol dose for the associative measures could be attributed to the simplicity of some of the associative tasks; but compare the biphasic hypothesis for the association findings.

Considering the relatedness of the task complexity and disinhibition hypotheses, it should be no great surprise that task complexity, while a good first working hypothesis, in itself hardly provides an accurate description of the effects of c.n.s. depressants on psychological functioning.

Implicit Versus Explicit Tasks: Related to the hypothesis of task complexity is that formulated by Payne, Moore, and Bethurum (1952), who studied the effects of depressants (motion sickness preventatives) on various tests of known factor content. These investigators observed a "hierarchy of effect" similar in kind for the several depressant agents studied, ranging "from psychomotor coordination, with no demonstrable effect, through routinized linguistic and perceptual process with moderate effects, to intellective processes in which the operations are highly implicit and somewhat disengaged from the immediate and continuous control of external stimuli." These investigators concluded that the "drugs exerted uniformly the most pronounced effects upon those psychological functions which are highly implicit, such as visual imagery, ideation, mental set, and judgment . . . [and] the least pronounced effects upon those functions which are most explicit, such as perceptual-motor and spatial decisions activities."

Although highly suited to the findings of Payne, Moore, and Bethurum, this explanation provides no more accurate an account of depressant effects in general than does the hypothesis of task complexity.

Task Familiarity: Task familiarity was suggested by Jellinek and McFarland as a hypothesis subsidiary to that of task complexity. "It would thus seem that familiar tasks are less affected by alcohol than the unfamiliar, even if the latter are less complex." "On the whole, it would seem that tasks of dexterity and skill are affected by alcohol in proportion to their complexity but that familiarity with the task may compensate for, and even outweigh, the element of complexity unless the latter is of a very high degree as in typewriting. Within a task of given complexity, familiarity with the task tends to lower the effect of alcohol."

A related hypothesis, "that retention of a serial order verbal task [is] rendered substantially less liable to drug impairment when original learning [is] carried beyond the point of sheer mastery," was tested and confirmed[1] for several motion sickness preventatives by Payne and Hauty (1957).

The hypothesis of task familiarity is a bit difficult to evaluate by itself. The combined complexity-familiarity hypothesis seems quite plausible, yet the present study provides striking exceptions: Abstract Reasoning (4) and Digits Backward (3), tasks which were unfamiliar as well as complex, were neither of them markedly impaired by the heavy alcohol dose. Considerations of complexity and familiarity, even when combined, provide no more than a good first working hypothesis.

The Role of Attention: That cancellation appeared to be more impaired by alcohol than tachistoscopic measures of perception was attributed by Jellinek and McFarland to the more active attention required by the cancellation measures.

Drug actions, particularly those of depressant agents, may also be revealed by task aspects which do not fully engage Ss' attention. Consider a depressant drug that acts to slow down Ss' writing. Asked to maximize writing output, resolute Ss may manage to maintain their writing pace, drawing on reserve capacities to compensate for the drug-induced retardation, perhaps modifying their usual writing style in the process. Even though the drug's presence may not be revealed by changes in writing speed, it may yet be betrayed by changes in Ss' characteristic modes of expression.

Ss asked to perform a given task are, of course, unlikely to be conscious of the entire range of means available to help them attain the desired end. By observing details of Ss' procedure not too obviously linked with the prescribed task, the investigator may discover altered behaviors that have escaped the attention

---

[1] Compare the present (non-significant) finding that Easy Associations were less impaired by the heavy alcohol dose than Hard Associations. See the section: *Recall*, Chapter 6.

of Ss in single-minded pursuit of their prescribed objective. The investigator may thus observe drug effects on horizontal writing extent that are perhaps a by-product of Ss' efforts to compensate for the drug's tendency to retard writing pace.

Detection of drug effects might therefore be enhanced by the following stratagem: Distract Ss by asking them to maximize output of a particular kind. Then observe aspects of performance incidental to the task set for Ss, aspects to which Ss themselves pay little attention. Since drug-induced changes in such incidental measures do not lead to efforts at compensation, one difficulty in the way of detecting drug effects is obviated.

Interesting and practical in its implications as this stratagem may be, it is after all merely based on an *ad hoc* hypothesis formulated to account for the Writing Expansiveness (1) findings. Even these findings can be explained in other (cerebellar) terms, without reference to incidental measures. The incidental task features hypothesis thus remains simply an interesting suggestion, worthy of further exploration.[2] (Note the related hypothesis arrived at independently by McFarland [1939], to the effect that habits beyond conscious control [such as eye movements] may be especially sensitive to the effects of oxygen deprivation.)

Test Length: That brief tests, even when reliable, may fail to reveal impairment has frequently been observed. "Disordered persons do not," according to Zubin (1948), "differ markedly from normal individuals . . . as long as the task is short . . ." Dunlap (1918) noted that "under asphyxiation, as under alcoholic intoxication, it is possible for [S] to 'pull himself together' for a brief span of time (a minute or even several minutes), during which his efficiency on a set task may be as high as in his normal condition . . . In this way, his real psychologic deterioration may be concealed [when tests are single and brief]." Similar observations were made by Drew, Colquhoun, and Long (1958) while studying alcohol effects on performance. The technique of continuous work was, indeed, originally introduced by Kraepelin "because

---

[2] Word Count (28) was enhanced, perhaps because of its "incidental" nature, both by caffeine and by the small alcohol dose. But the incidental task features hypothesis, here formulated in terms of *compensation* for impaired functioning, is not so readily applied to heightened functioning.

tasks involving but short intervals of performance failed to bring out the deficit evident in everyday life" (Hunt and Cofer, 1944).

Effects of stimulants may also be more readily demonstrable under conditions of task prolongation (Barmack, 1940; Seashore and Ivy, 1953; Tyler, 1947), for these drugs not only *improve* various kinds of performance, they may also *allay* impairment brought on by fatigue or other central factors. Task prolongation can thus confound two kinds of drug effects (this holds true for depressants as well as for stimulants), and is not recommended to the investigator who wishes to identify and to segregate a drug's effects on *particular* functions.

The present findings indicate how detection of drug effects may be affected by test length. When tests were so brief as to be unreliable, the likelihood of significant results was not too great. The time limit for Street Gestalt Completion (16) was especially brief—fifteen seconds; the correlation between post-drug and pre-drug values, over all four treatments, averaged +0.57. Street Gestalt Completion was unaffected by drugs. The time limit for Arithmetic Errors (9) was somewhat longer—ninety seconds; the correlation between post-drug and pre-drug values, over all four treatments, averaged +0.62. Significant findings were more nearly attained by Arithmetic Errors than by Street Gestalt Completion, but the Arithmetic Errors results would likely have been even more definite had the number of test items been doubled, for example, and had the time limit been, say, three minutes rather than a minute and a half.

It is apparent from the present findings that drug effects can be demonstrated by brief tests, provided the tests are not so brief as to be unreliable. Thus, drug effects were revealed for Free Association (14) (time limit: two minutes), Mutilated Words (25) (time limit: two minutes), Digit Symbol Composite 1 (17) (combining two one-minute trials), Clerical Speed and Accuracy: Part I (21) (time limit: three minutes), and Addition, Correct (8) (time limit: five minutes). The findings for incidental measures such as Writing Expansiveness (1) (based on two one-minute trials) also testify to the fact that drug effects can be detected by brief tests.

Considering the brief tests affected by drugs in the present study, and the relatively lengthy tests—Abstract Reasoning (4), Language Usage (26), Digits Forward (2), Digits Backward (3), for example—unaffected by drugs, one might actually be tempted to surmise that the longer the test, the less its susceptibility to drug effects. Of course, this generalization runs counter to the experience of some investigators. Besides, the longer tests of the present study were generally "power" tests, which are considered below.

A test's length does not then seem to be a crucial determinant of its sensitivity to drug effects, unless the test is so short as to be unreliable or so long as to affect Ss' task attitude and perhaps obscure the meaning of the test results.

Tasks Requiring Rapid Functioning: Least affected by drugs in the present study were such power tests as Abstract Reasoning (4) and Language Usage (26), which were usually completed within the allotted times. Most susceptible to drug effects were such speeded tests as Clerical Speed and Accuracy (21), Mutilated Words (25), Digit Symbol (17), Addition (8), Free Association (14), and Word Fluency (15).

Consider especially the tests of immediate, intentional recall. Though these tests were not all markedly affected by drugs, the Immediate, Intentional Recall Composite was significantly enhanced by caffeine and significantly, if less markedly, impaired by the heavy alcohol dose. *The only measure in the present study significantly affected both by caffeine and by the heavy alcohol dose*, the Immediate, Intentional Recall Composite was certainly a sensitive indicator of drug effects.

The immediate, intentional recall tests, while not speeded in the ordinary sense of the term, at least constituted a special group of power tests. Unlike true speeded tests, these recall tests did not explicitly require Ss to minimize response time. Successful performance did, however, call for rapidity of functioning. Information was delivered rapidly to Ss, who had to receive and process the information as it arrived while storing information received earlier, all without appreciable distortion of the form or content of the information received and stored. An indirect premium was thus placed on reduced response time, for the longer the apprehended information was held in mind, the greater the likelihood that specific items of information would be forgotten and that the original order of presentation would be scrambled on reproduction.

Sensitivity to drug effects appears then to vary directly with the rapidity of functioning required by the task. This might perhaps be due to drug-induced alterations of temporal processes in the central nervous system. Though fairly well suited to the present results, this "temporal hypothesis" is yet not fully adequate in accounting for other studies' results.

Clearly, response to classical c.n.s. agents such as alcohol and caffeine may be affected by a number of task features, no one of which appears to be of crucial importance for drug investigations. Nor does any single hypothesis or combination of hypotheses as yet appear to provide a clear understanding of the *means* by which drugs alter higher mental functions. Such understanding may yet be provided by investigations of drugs and doses whose psychological effects are mediated primarily by the highest levels of c.n.s functioning, and little, if at all, by lower c.n.s. levels or by sites outside the central nervous system.

PART IV
Conclusion

12

SUMMARY AND
CONCLUSIONS

FIFTY-SIX NORMAL, young adult volunteers received one of four doses (each containing grape juice, water, and peppermint oil), assigned according to a randomized blocks design:

1. Ethyl alcohol, 15.7 ml. absolute per square meter of body surface area—approximately equivalent to the alcohol content of two martinis. Maintenance doses enabled blood alcohol levels to be kept within a range of about 15 to 60 mg. per 100 ml. of blood.

2. Ethyl alcohol, 31.4 ml. absolute per square meter of body surface area—approximately equivalent to the alcohol content of four martinis. Maintenance doses enabled blood alcohol levels to be kept within a range of about 45 to 90 mg. per 100 ml. of blood.

3. Caffeine alkaloid, 100 mg. per square meter of body surface area—approximately equivalent to the caffeine content of two cups of coffee. Maintenance doses enabled blood caffeine levels to be kept within a range of about 0.5 to 5.0 mg. per liter of blood.

4. A placebo. The dose given to placebo Ss contained neither alcohol nor caffeine.

A large variety of psychological tests was administered, pre-drug; these tests were repeated, with some variations, post-drug. Internal drug levels were determined from blood samples obtained by finger puncture. Identical procedures were followed, both for test administration and for finger puncture, irrespective of the dose assigned to S.

Caffeine generally acted to *mobilize* the intellectual resources of the organism. These resources were generally *immobilized* by the larger alcohol dose; direct c.n.s. effects of this dose were confounded with side effects. The effects of the smaller alcohol dose were generally negligible. These over-all drug findings are in accord with those reported by previous investigators.

The effects of a given dose were not entirely uniform, but varied from one psychological function to another. Two psychological actions of caffeine were particularly marked:

1. Caffeine acted most definitely to increase spontaneity of association. The process of thought formation was quickened, and associations became more abundant. Ss were less likely to be at a loss for ideas or for words. Addition, a mechanized task involving association along logical lines, was facilitated.

2. Immediate, intentional recall of a rapid influx of auditory information was equally enhanced by caffeine. Ss appeared better able to take advantage of the incoming information, organizing it so that it could be better assimilated and formed into a stable memory trace.

While occasional unfavorable caffeine effects were noted in the present study, these effects in no instance attained nor even approached significance. That caffeine actions are *not uniformly beneficial* is, however, clear from *previous* reports relating to hand steadiness. It is perhaps ironic that the effect of caffeine on hand steadiness, perhaps the most definitely established psychological (or near-psychological) caffeine effect, is one of *impairment*. Caffeine action thus differs qualitatively (as well as quantitatively) from that of the amphetamines, therapeutic doses of which appear to benefit all psychological (and near-psychological) functions, *including* hand steadiness.

The action of the heavy alcohol dose, which was not simply antagonistic to that of caffeine, was characterized by several distinctive features:

1. Markedly impaired by the heavy alcohol dose were performances requiring Ss to take in visual details at a glance, to discriminate rapidly among those details, and to make sense out of meaningful visual patterns which had previously been disrupted. These impaired performances were very likely mediated by drug-induced disturbances of visual acuity, perceptual closure, and coordination of eye movements.

2. Recall of impressions that had just been committed to memo-

ry was also clearly impaired by the larger alcohol dose. The disturbance of recall and of visual thinking accounted in large part for the marked disruption of performance on a complex task (Digit Symbol) involving both eye-hand co-ordination and learning of rote associations.

3. Horizontal writing extent was markedly increased, presumably due to cerebellar impairment produced by the larger alcohol dose. Drug effects for writing extent greatly exceeded those for writing pace. Depressant action may, then, be more readily revealed—what with Ss concentrating their energy reserves on prescribed aspects of a capacity task to compensate for drug-induced impairment—by involuntary than by deliberately adaptive aspects of capacity tasks.

The smaller alcohol dose tended to *facilitate* association, but produced no other psychological changes worthy of notice.

Only Ss receiving the heavy alcohol dose were aware of being markedly affected by the dose administered to them. The examiner, who was not informed of the doses given to individual Ss, was rather well able to discern from Ss' gross (non-test) behavior which Ss had received the larger alcohol dose. Findings for the larger alcohol dose were therefore obtained under conditions which could hardly be termed "double-blind."

The examiner was essentially unable to distinguish Ss receiving the other three doses. The examiner failed to distinguish the effects of caffeine and placebo; furthermore, caffeine Ss hardly suspected that a stimulant had been administered. The caffeine findings were thus obtained under virtually ideal double-blind conditions.

The double-blind procedure is seen to be effective under some experimental conditions, and with some but not all dosage forms. It is a procedure with definite potentialities and limitations. Routine evaluation of the efficacy of double-blind procedures is recommended for *all* experiments employing such procedures, in order that the validity of an experiment's results might be better assessed and the meaning of these results better understood.

Of greater theoretical significance are two further problem areas which, while emerging from our consideration of alcohol and caffeine effects, have implications for the study of *all* classical depressant and stimulant drugs.

Consider first problems concerning the separation of the effects of capacity and attitude. Highly motivated Ss, sensing an intellectual deficit induced by depressant drugs, may be able to compensate temporarily for this deficit, thereby obscuring the drugs' primary effects. Although compensation for deficit is said to play an important role in studies of depressant drugs, a clear demonstration of such compensation, and an estimate of its magnitude, would be desirable.

Confounding of attitude and capacity effects is also a problem in studies of stimulant drugs, whose beneficial effects on performance could be mediated by a drug-induced enhancement of capacity, or attitude, or both. The evidence strongly suggests that stimulants such as caffeine can indeed enhance capacity directly. This conclusion is not final, however, for stimulant effects on capacity are not easy to disentangle from stimulant effects on attitude. Here is a problem that merits further study.

Perhaps more significant for an understanding of the psychological actions of classical depressants and stimulants is yet another set of problems—concerning the dimensionality of drug effects on the intellect, and the hypothesized release of lower c.n.s. centers from inhibition by higher centers—posed by present as well as by previous findings for alcohol and caffeine. Consider the following questions:

1. Are the effects of classical depressant and stimulant drugs describable simply in terms of enhancement or impairment of specified functions? Do depressants merely impair, and stimulants merely enhance, given modes of psychological functioning? Is the relation between actions of these two drug classes one of unqualified contrast?

2. Are the effects of classical depressants and stimulants a monotonic function of dosage level, intensity of drug effect increasing with dosage level? May these effects be biphasic, *direction* of effect being reversed as dosage level increases?

3. Does a given depressant or stimulant have equivalent effects on a variety of psychological functions? Are the intellectual effects of such a drug *unidimensional*? Is there an interaction between drugs and intellectual functions, with profiles of drug effect varying from drug to drug?

Summary and Conclusions

These are questions of fact. They do not necessarily refer to mechanisms of action. But questions relating to depressant drugs have traditionally been answered by appeal to the Jacksonian scheme of dissolution of nervous function. According to the traditional view, depressant drugs progressively immobilize higher nervous centers, the locus of disturbance descending (irregularly) along the cerebrospinal axis of the central nervous system. The traditional view further implies that all effects of a depressant drug must be regarded as phenomena of impairment brought about by disinhibition of lower nervous centers.

The disinhibition hypothesis is more than merely plausible. It has actually proved most useful in ordering the effects of a wide range of doses of depressant drugs. To have enlarged our understanding of the stages of anesthesia is in itself an impressive achievement of this hypothesis. Despite its fruitfulness, however, the disinhibition hypothesis approaches its limits of usefulness when applied to the psychological effects of *mild* doses of depressant drugs. Some performances as well as some functions do appear to be *enhanced* by mild doses of *depressants*—despite the contradiction in terms. That depressants have enhanced *some* functions has often been overlooked or explained away in an attempt to arrive at a clear and simple picture of the facts. But psychological enhancement is rendered no less real by a hypothesis that translates it into physiological impairment.

The strong emotions sometimes aroused by the disinhibition hypothesis have resulted in a tendency to foreclose discussion about the enhancement of function by depressant drugs. Such foreclosure is clearly premature.

To think of depressant drugs as acting simply along the vertical axis of the central nervous system, and not to consider the possible actions of these drugs along non-vertical dimensions of the central nervous system, is to disregard the empirical findings. The observed interactions between drugs and psychological functions suggest an alternative to, or rather modification of, the disinhibition hypothesis: Depressant drugs alter the balance among *coordinate* c.n.s. centers as well as among centers located at different levels of the cerebrospinal axis.

Appendices

APPENDIX A

THE DOUBLE-BLIND PROCEDURE: RATIONALE AND EMPIRICAL EVALUATION[1]

PROBLEM: Suggestion is a process that may readily modify the effects of physical treatments administered to human Ss. No one is more aware of this problem than the physician, who has traditionally assigned sugar pills or pink water to patients needing reassurance by some material token of the physician's continuing concern for their welfare. Administration of a placebo (an inert substance bearing a superficial resemblance to an active medication, designed to "please" rather than to provide physical benefit for the patient) has come to play an important role not only in clinical practice, but also, as one aspect of the "double-blind" procedure, in scientific experimentation. The by now classical double-blind procedure, which aims to blind both S and examiner to selected features of an experiment, represents a significant attempt to minimize suggestion effects.

The double-blind procedure has, indeed, come to be regarded by many as the very mark of a "scientific" experiment on drug effects. But acclaim for the double-blind procedure is by no means universal. While esteemed by laboratory investigators concerned with determining the nature of drug effects and with estimating drug potency, this method tends to be viewed with skepticism by investigators having more direct responsibility and concern for the welfare of individual patients. Thus, clinical investigators

---

[1] Condensed and reprinted, with permission, from *J. nerv. ment. Dis.* (Nash, in press).

133

are deprived by the double-blind procedure of information considered useful if not essential for the judgments formed in everyday medical practice. Clinicians may therefore regard the double-blind procedure as impractical, irrelevant, unwarranted, and perhaps a bit inhuman: impractical because incapable of realization; irrelevant and unwarranted because it intrudes into normal medical practice and impedes the physician's efforts to provide optimum medical care for his patients; inhuman because it requires a control treatment, in which patients are usually deprived of active medication.

The double-blind procedure is thus often viewed with more than mere skepticism. Opposition to this procedure has become but one aspect of a more general antagonism to rigorous experimental controls—including an antagonism to such prerequisites for unbiased experimentation as the strictly random assignment of Ss to experimental and control groups, and even an antagonism to the very use of control groups (cf. Nash, 1959). In this emotionally charged atmosphere, in which the double-blind procedure has become a focal point of disagreement between laboratory scientist and intuitive clinician, attitudes have unfortunately polarized.

How resolve these conflicting attitudes regarding the double-blind procedure's practicability and desirability? We are clearly in need of guideposts to help us determine the double-blind procedure's true status.

Suggestion: Suggestion may alter an experiment's outcome by increasing experimental error or (and this is by far the more serious disturbance) by biasing experimental results (Nash, 1959). Let us examine some of the principal means by which suggestion distorts the results of experiments on the psychological effects of drugs or other forms of physical treatment.

Consider how experimental findings may be biased by the investigator's attitudes. The investigator may have a favorite hypothesis to which he clings, that he wishes to see confirmed. He may on the other hand desire to obtain *significant* findings, being relatively unconcerned about the kind of findings obtained. In either instance, the investigator's self-regard and professional prestige may become involved in the study's outcome.

While some examiners undoubtedly "bend over backwards" to avoid inducing S to behave in a desired fashion, spurious positive results do tend to be favored by an examiner's knowledge of the physical treatments administered to particular Ss. The examiner may be tempted, consciously or otherwise, to alter the general atmosphere of the examination, or the manner of presentation of test procedures, or the quantification of subjective impressions—each of these changes provides an opportunity to introduce bias into the test results. Thus, an examiner wishing to demonstrate a drug-induced acceleration of response may be aware neither of his rapid and vigorous delivery of test instructions nor of the extent to which Ss are alerted by his forceful manner.

Experimental results may also be biased as a result of attitudes (conscious or unconscious) held by S. Holding medical research, and science generally, in high esteem, S likely wishes the investigation success. This would be especially true for normal volunteer Ss and for patients suffering from a condition whose alleviation is being investigated. Invested with the authority and glamor of science, the examiner tends to command the respect of his Ss, who are then motivated by their personal relationship to the investigator as well as by a desire to contribute to medical and scientific progress. Although Ss may come to oppose the examiner's wishes during the course of the experiment, they more likely desire to please him. Wishes to forward rather than impede the supposed aims of the investigation and of the investigator, wishes to comply with rather than resist the supposed effects of the treatments being investigated—such wishes, however subtle, are likely to favor spurious positive results.

Features of the Double-Blind Procedure: These, then, are the kinds of error against which we desire to be safeguarded. Bias due to suggestion may be minimized by insuring identical, or at least near-identical, conditions of experimentation (except for the particular treatments being investigated) for experimental and control Ss. A variety of measures has been proposed to minimize and if possible eliminate completely systematic errors arising from the suggestibility of examiners and Ss.

First, consider the *subject*. The principal means of controlling his suggestibility is to keep S ignorant of, or blind to, the experiment's actual nature. The "experimentally naive" person is the ideal S.

Measures of greater or lesser severity may be adopted to prevent Ss from learning about the experiment's basic purpose. Ideally, one would avoid announcing which active medications are being studied. One would furthermore withhold from Ss the fact that more than a single treatment is being administered. Should Ss suspect that two or more dosage forms are included in a given study, they should by no means be informed of the specific treatments assigned at particular times or to particular Ss.

Such withholding of information is most feasible with captive Ss, such as hospitalized patients. Greater skill and care is required in withholding information from volunteer Ss: steps should be taken to maintain rapport with S, whose desire to co-operate (a basic prerequisite for many but not all experiments) might otherwise be impaired by S's uncertainty regarding the nature of the medications administered. Unless Ss have knowingly volunteered for an experiment having possibly harmful effects, they need to be reassured that no harm will come them.

Should Ss remain unaware of the treatments they receive, their suggestibility will not bias the experiment. Being blind to the true nature of the investigation does not of course prevent S from speculating about the investigation. Such speculation may in fact be stimulated if S believes that information is being withheld from him. But suggestibility effects arising from blind speculation tend to be distributed among the various treatments included in a given study. "The enthusiastic hope of the patient and the investigator for successful results from the drug," noted Lasagna (1955), "must be bridled, or at least allowed to distribute itself equally between the drug and the placebo." The suggestibility of Ss unable to distinguish experimental from control treatments may inflate experimental error, but it does not introduce bias into the experiment.

The mere withholding of information about the experiment's design is not in itself sufficient to blind S to the facts of the experiment. Should S be examined now under *untreated* control conditions, now under the influence of an active medication, he can certainly distinguish the two conditions, though he may be

Appendix A

unable to identify the precise medication employed. Should S be examined under but a single treatment condition, he may remain unaware that other treatment conditions are being employed; but each S may yet learn about the structure of the experiment, given an opportunity to share experiences with other Ss.

It is for these reasons that the placebo has been introduced into drug research. For greatest effectiveness, oral placebos used for experimental purposes are made up to duplicate or at least simulate the taste, smell, color, and superficial form of the active medication. A placebo is administered to control Ss in order to encourage their belief that they are in fact receiving the active medication. Thus, the investigator not only *promotes ignorance,* he also *engages in active deception* in an effort to control the effects of suggestion.

Further precautions may yet be advisable. Though having its primary effects on, say, the liver, which functions "silently" (i.e., without subjective concomitants), an active medication may still be betrayed by subjective symptoms absent with the inert placebo. Cues to S may be minimized under such circumstances by employing an "active" or "positive" placebo, having subjective effects mimetic of the active medication's "give-away" effects. (Alternatively, placebo and active medication may each be administered in a vehicle that simulates or masks superficial subjective effects of the latter.) An active placebo is useful only when the "give-away" subjective effects of the corresponding active medication are not themselves the subject of investigation.

Numerous as are the above precautions, they may still be insufficient to forestall bias due to subject suggestibility. Ss having some familiarity with medical experiments may anticipate and therefore more readily suspect the use of placebo controls, in the absence of expected drug effects. And previous experience with drugs may enable Ss to recognize active medications, by their subjective effects; to detect inert placebos, by the absence of familiar subjective effects; and to entertain doubts about active placebos, because of their imperfect simulation of the active medications' subjective effects. The double-blind procedure's effectiveness may then be buttressed by careful selection of Ss; Ss familiar neither with the active medications nor with the problems of experimental design merit the highest priority.

Consider now measures designed to establish and maintain *examiner blindness*, in order to bring about an equal (or more nearly equal) distribution of suggestion effects between experimental and control conditions. The primary objective is to keep the examiner ignorant of selected features of the experiment. The *investigator* is naturally aware of the various active drug and control treatments employed in a study. It is unlikely, though not inconceivable, that an *examiner* who is not himself the primary investigator would forego such knowledge.

Because the examiner's knowledge of the experiment so greatly exceeds that of S, it becomes imperative to exercise strict control over whatever knowledge can be denied the examiner. Efforts are therefore concentrated on procedures for withholding from the examiner information about the particular treatments assigned to individual Ss. Such procedures have undergone some institutionalization.

Thus, the examiner absents himself while Ss are randomly assigned to treatments; while doses are prepared for individual Ss; and (for non-identical appearing doses or for doses producing distinctive sensations on ingestion) while doses are being administered. Test protocols whose scoring requires quantification of subjective impressions are only scored after completion of the examination, without knowledge of the treatments assigned to individual Ss. These protocols are frequently coded before scoring in order to mask Ss' identities, or they are processed by scorers unfamiliar with the treatments assigned to particular Ss.

The adoption of standardized or semi-standardized procedures for regulating the blindness of S and examiner does not of course imply any reduction of the examiner's responsibility for the conduct of the examination. The examiner ought still to strive for the utmost objectivity in his conduct of specific examination procedures, he ought still to do his best to avoid contaminating the examination's atmosphere by his personal wishes regarding the experiment's outcome. But examiners vary in their self-critical capacity, and even the best examiner's self-knowledge is limited. It is just because of the examiner's inability to insure his complete objectivity that rigorous procedures have been designed to supplement, *but not to supplant,* the examiner's own efforts.

An Empirical Evaluation of Double-Blind Procedures: These then are the techniques that, combined in any of a variety of ways, are collectively termed "the double-blind procedure," and this, the rationale underlying this procedure. How adequate is this rationale, how well does the theory work out in practice?

The double-blind procedure's effectiveness is of course open to study, and an empirical evaluation of the procedure would indeed be desirable. Consider, for example, the effectiveness of techniques adopted in the present study for the purpose of establishing and maintaining the blindness of examiner and Ss.

1) *Procedures of the Present Study:* Various procedures of the present study were designed to withhold knowledge, from examiner and Ss, that could lead to bias due to suggestion.

Thus, the investigation was advertised simply as a medical study. Prospective Ss, inquiring for further details, were informed that they would be expected to imbibe alcoholic beverages, undergo psychological examination, donate very small quantities of blood, and abstain from a few meals. Although Ss were aware that the alcoholic beverages were more than an incidental feature of the experiment, the examination was conducted in an atmosphere designed to obscure the experiment's actual purpose.

The investigation was described to prospective Ss in a manner calculated to foster their belief that each S received an alcoholic beverage. Ss were not informed that *several* treatments were included in the experiment. Nor was mention made of the fact that Ss might receive either of two strengths of alcohol; or some active medication other than alcohol; or no active medication at all. Care was taken to avoid suggesting that beverages varied from S to S according to body surface area or, indeed, on any basis whatever. Least of all was the particular experimental treatment assigned to S disclosed to him.

These measures, undertaken to maintain Ss' ignorance of the experiment's true nature while fostering each S's belief that his beverage contained alcohol, would not in themselves have been sufficient to convince caffeine and placebo Ss that they were in fact imbibing alcohol. Some more positive measure was required to enable caffeine and placebo Ss to accept the idea that their drinks were indeed alcoholic. To this end, Mead's (1939) procedure was employed, in which *all* doses, alcoholic or otherwise, were administered in a grape juice-and-water vehicle flavored with peppermint oil—a pungent substance whose immediate sensory effects, on contact with the membranes of mouth and stomach, resemble those of alcohol. The peppermint oil was intended to mask the absence of ethyl alcohol (at least during the interval before Ss would expect to feel alcohol's full effects) in drinks supplied to caffeine and placebo Ss. This deception was aided by the (somewhat accidental) fact that most Ss were rather light drinkers; about one-fifth of the Ss completely abstained from alcohol or drank but a glass or two of wine per year, and were thus somewhat naive about alcohol effects.

Subject blindness was perhaps further promoted by the fact that Ss were examined singly, post-drug. By avoiding giving Ss an opportunity to learn about variations in dose (even if only on the basis of body surface area) from S to S, it was hoped to discourage Ss from giving undue thought to, and perhaps correctly surmising, the content of their beverages.

Steps were taken to maximize the blindness not only of Ss, but also of the examiner. The examiner was the investigator himself, who was naturally aware of the doses employed in the study. He was not, however, informed of the particular treatments assigned to individual Ss—such assignment was made by an assistant, who also prepared and administered the beverages. While the examiner could not help but notice Ss' spontaneous behavior—their gross, non-test behavior (including some comments by heavily alcoholized Ss suggesting that their functioning had been impaired)—he avoided close questioning or careful observation of Ss, except as required by the examination. The examiner avoided trying to discover, by direct or indirect means, which treatments had been assigned to particular Ss. Ss' test responses were mostly recorded without being seen by the examiner, who made no attempt (during the examination) to compare Ss' pre-drug and post-drug responses in order to determine whether performance had improved or deteriorated following treatment. Tests requiring

Appendix A

TABLE 21
ACCURACY OF EXAMINER'S GUESSES REGARDING TREATMENTS ASSIGNED TO Ss

Treatments Guessed by Examiner	\multicolumn{4}{c}{Treatments Actually Assigned to Ss}
	CAFFEINE	PLACEBO	ALCOHOL (15.7 ML./M.^{2})	ALCOHOL (31.4 ML./M.^{2})
First treatment session:				
Caffeine	3	3	3	0
Placebo	11	10	6	1
Alcohol (15.7 ml./m.^{2})	0	1	4	3
Alcohol (31.4 ml./m.^{2})	0	0	0	10
Second treatment session:				
Caffeine	2	1	1	0
Placebo	7	10	6	1
Alcohol (15.7 ml./m.^{2})	3	2	7	2
Alcohol (31.4 ml./m.^{2})	0	0	0	8

subjective evaluation[2] were scored after all Ss had been examined, without knowledge of Ss' identity; test protocols were coded before being scored by the examiner, or they were scored by persons unfamiliar with the Ss.

2) *Evaluation of Examiner Blindness:* These were the procedures adopted in the hope of approximating double-blind conditions. Just how successful were these procedures? What can be said about the *realization* of double-blind conditions in this study?

First consider examiner blindness. As each of S's last two (i.e., post-drug) sessions ended, the examiner guessed which treatment S had received. These guesses are summarized in Table 21; results for each treatment session are tabulated separately. (Note that fewer judgments were made for the fourth session than for the third, because of the heavily alcoholized Ss unable to participate in the final session.) The results indicate that, despite the measures adopted to insure examiner blindness, the examiner was somewhat able to discern which treatments had been administered to individual Ss.

In evaluating the data of Table 21, bear in mind that the exam-

---

[2] Writing Intensity (4, 14), Sentence Memory (5), Consequences (7), Controlled Associations (10), Plot Titles (11), Street Gestalt Completion (16), Story Memory (20, 23), Contingencies (24), Sentence Completion (28), and Rosenzweig Picture-Frustration (32).

iner was moderately successful in preserving the independence of his judgments regarding treatments assigned to Ss, notwithstanding two factors tending to prejudice his thinking:

a) The examiner strove to judge S's condition on its own merits. The examiner knew, for example, that Ss had been admitted to the experiment in consecutive order; that successive groups of four Ss had each been designated a "block"; and that the four treatments had been assigned separately to the Ss of each individual block. In judging the condition of a given S, then, the examiner did his best to forget his judgments about other Ss belonging to the same block. The examiner went even further to keep his judgments from being undermined by spurious factors: he assumed that each S had been treated with placebo, in the absence of positive evidence pointing to one of the active treatments. That the examiner succeeded, at least in part, in ignoring the "blocks" aspect of the experiment is indicated by the fact that the number of Ss thought to have received placebo greatly exceeded (rather than being equal to) the number of Ss thought to have received caffeine.

b) The examiner also managed to keep his second-treatment-session guesses about given Ss from being overly influenced by his first-treatment-session guesses about these Ss. Note the shift, between first and second treatment sessions, in the number of Ss assumed to have received the smaller alcohol dose, and in the number assumed to have received caffeine.

The examiner's apparent reluctance to attribute Ss' post-treatment conditions to caffeine is understandable, considering his inability to distinguish caffeine Ss from placebo Ss.

Other treatments were distinguished by the examiner with varying degrees of success. The smaller alcohol dose was distinguished from placebo and caffeine, combined, with only borderline success ($p = 0.05$, for the first treatment session; not significant, for the second treatment session [Armsen, 1955]). The larger alcohol dose was readily distinguished from the smaller alcohol dose ($p = 0.05$ and $p = 0.01$, for the first and second treatment sessions, respectively); from placebo and caffeine, combined ($p = 0.01$, at the very least, for each treatment session); and from placebo, caffeine, and the smaller alcohol dose, all combined ($p = 0.01$, at the very least, for each treatment session). Although recognition of the heavily alcoholized Ss was facilitated by the fact that six of them became ill and were unable to complete one or both post-drug sessions, the intoxicated state of these Ss was, in most instances, apparent long before they became ill. At any rate, inclusion of the indisposed Ss (three dropped out during the first treatment session; three others, during the second treatment session) does not substantially alter the results for either treatment session.

3) *Evaluation of Subject Blindness:* Consider now subject blindness, of which there are several indices.

There are, first of all, the examiner's qualitative observations: Ss gave no indication of suspecting that caffeine was included in the experiment. But one caffeine S did express his surprise at feeling fresher, not at all the reaction he had expected with alcohol.

Nor did Ss appear to suspect that the alcoholic content of their drinks varied from S to S. Although Ss held various opinions regarding the alcohol content of the beverages each had drunk, all but two Ss appeared to believe that their drinks contained some alcohol.

This latter impression of the examiner is supported by a quantitative analysis of Ss' responses. At experiment's end, Ss were asked: "How much whiskey would you have to drink at once to feel like you did during the session just completed?" One-tailed *t*-tests for Whiskey Estimate, Corrected[3] definitely indicate that both caffeine and placebo Ss thought the whiskey-equivalent of their beverages exceeded zero ($p = 0.0005$, at least, for caffeine; $p = 0.001$, for placebo).

A few of the 28 Ss who received no alcohol did comment on the surprisingly low alcohol content of their beverages, and two Ss wondered aloud whether their beverages had contained alcohol.

---

[3] Since many Ss were uncertain about the effects typically produced by given quantities of whiskey, a corrected whiskey estimate was computed by taking into account S's estimate of the amount of whiskey required to lay him low (see Whiskey Estimate, Corrected [35], Chapter 3). *t*-tests are reported above for the corrected but not for the uncorrected estimate, since data were distributed normally only for the corrected measure.

While these two Ss suspected they had received no alcohol, they were reluctant to make an outright assertion to this effect. Even though several Ss reported Whiskey Estimates (35) that were negligible, *all* Ss provided estimates exceeding zero.

Although Ss failing to receive alcohol did not notice, or would not assert, the complete absence of alcohol from their drinks, Whiskey Estimates did reflect the differences in the alcohol content of drinks assigned the four treatment groups (see Table 18). Whiskey Estimates of Ss administered the larger alcohol dose exceeded[4] those of Ss administered the smaller alcohol dose, as would be expected both from the greater fluid volume ingested by Ss receiving the larger dose and from the marked effects of this dose. Yet Whiskey Estimates even discriminated treatment groups imbibing comparable fluid volumes and undergoing comparable treatment effects on performance. Thus, Whiskey Estimates reported by the lightly alcoholized Ss exceeded those for placebo Ss. It is clear that Ss were only partially blind.

While being examined, most Ss expressed curiosity about, but had not yet formed too definite an impression of, the beverage's composition. Ss' Whiskey Estimates, crystallized in response to the examiner's inquiry *at experiment's end,* therefore provided only limited information about subject blindness.

Additional evidence relating to Ss' blindness indicates that they unequivocally distinguished neither treatment effects on performance nor specific subjective treatment effects. Thus, when S was asked, also at experiment's end, to compare his pre-drug and post-drug performances on each of five tests, Ss under the influence of the heavy alcohol dose correctly reported the greatest impairment of performance, but none of the active treatments *significantly* affected Estimation of Treatment Effects (34) (see Table 18).

Furthermore, the active treatments produced no very marked effects on Ss' self-ratings (see Table 4). Caffeine tended to alleviate feelings of fatigue, but otherwise produced no noticeable subjective effect. The smaller alcohol dose had no significant effect on self-ratings. Only the heavy alcohol dose Ss, who experienced

---

[4] Were it not for the lack of Whiskey Estimates for the six incapacitated Ss, differences between Ss receiving the larger and the smaller alcohol doses would likely have been highly significant.

Appendix A

TABLE 22
Ss Volunteering for Future Studies

Status of Ss	Treatments Received by Ss			
	CAFFEINE	PLACEBO	ALCOHOL (15.7 ML./M.^{2})	ALCOHOL (31.4 ML./M.^{2})
Volunteers	9	6	8	1
Non-volunteers	5	8	6	13

an impairment of judgment and attention following treatment, provided subjective reports definitely altered by treatment.

Still further evidence suggests that Ss viewed the heavy alcohol dose in a different light from the other three doses (see Table 22). Given an opportunity to volunteer for further studies following the present one, Ss responded in surprisingly large numbers, except for those who had received the heavy alcohol dose. The distinction between heavy alcohol dose Ss and those receiving the remaining three treatments, combined, was significant at the 0.01 level (Armsen, 1955), even excluding the six Ss eliminated before the experiment's completion (see Table 23). (The one heavy alcohol dose S who did volunteer was, surprisingly, one of the six incapacitated Ss.)

Ss' adventurous spirits were certainly more dampened by the heavy alcohol dose than by the other doses.

The accumulated evidence indicates quite definitely that both examiner and Ss were fairly blind throughout the experiment, except in respect to the larger alcohol dose. Comparison between the large alcohol dose and placebo involved serious deviations from ideal double-blind conditions that could conceivably have

TABLE 23
Ss Volunteering for Future Studies
(Excluding Six Eliminated Ss)

Status of Ss	Treatments Received by Ss	
	POOLED PLACEBO, CAFFEINE, AND ALCOHOL (15.7 ML./M.^{2})	ALCOHOL (31.4 ML./M.^{2})
Volunteers	23	0
Non-volunteers	19	8

contributed to the impairment of performance associated with the heavy alcohol dose. Heavy alcohol dose findings might have been biased, as for example by lack of examiner blindness. Wishing to insure that the condition of obviously intoxicated Ss be reflected in statistically significant findings of impairment, the examiner might have altered (perhaps unconsciously) the atmosphere in which the examination was conducted. A slowing of the rate at which test instructions were presented to the heavy alcohol dose Ss might have reduced their response-readiness, and consequently their speed of response. (Despite possible exaggeration of heavy alcohol dose-induced impairment due to deviations from double-blind conditions, there is no question in the investigator's mind but that the heavier alcohol dose did indeed disturb performance.)

In contrast to the comparison between the heavy alcohol dose and placebo, that between caffeine and placebo appears to have been accomplished under virtually ideal double-blind conditions.

Conclusion: Having examined the double-blind procedure in actual practice,[5] what conclusions can we draw about the procedure's underlying rationale?

Let us note, first of all, that the procedure in question represents an attempt to apply the classical methods of experimental inquiry, developed by Bacon and Mill (cf. Cohen and Nagel, 1934), to situations readily obscured by suggestibility effects. The double-blind procedure refers to an assemblage of precautionary measures whose ultimate aim is the elimination of bias due to suggestion.

But application of whatever measures constitute *the* double-blind procedure in any given situation does not automatically bring about the desired conditions. It is therefore important to distinguish the procedures employed from the conditions realized. For the extent to which ideal double-blind conditions are approximated varies from situation to situation, even with the application of identical procedures. Parts of an experiment may be conducted in total or near-total blindness, while other parts of the same

---

[5] See Nash (in press) for a survey of previous, smaller-scale empirical evaluations of double-blind procedures employed in psychopharmacological investigations.

experiment may be completely lacking in blindness. One aspect of an experiment may be transparent to examiner and S alike, another aspect may be opaque to both. Alternatively (though this did not occur in the present study) an experiment designed to be double-blind may in fact only succeed in being single-blind, with either examiner or S learning about features of the experiment intended to be beyond their awareness.

Double-blind conditions are thus seen to be many-faceted and attainable in whole or in part. Double-blind conditions depend on a variety of factors, themselves often in interaction, and a given set of double-blind conditions may be brought about by different combinations of procedures. Ease of attainment of double-blind conditions is a function of the particular active medications and dosage levels being studied; of the routes of administration and dosage forms employed (Nash, 1959); of subjective effects (tastes, smells, burning sensations) induced at the moment of, or immediately following, ingestion of an oral dose; of subjective drug effects irrelevant to the experiment's purpose (nausea, for example), occuring after the drug's absorption; of the kinds of Ss employed and the relation between Ss and examiner; and so on.

Perhaps as crucial for the fulfillment of double-blind conditions is the investigator's understanding of the purposes that various double-blind techniques are intended to serve. Given such understanding, the experimenter is enabled to anticipate damage to the experiment by factors impeding attainment of double-blind conditions, and to adapt his procedures to the requirements of the occasion. The double-blind procedure is no rigid formula providing an "open sesame" for the solution of nature's problems. The investigator would be well advised to sharpen his understanding of the ways in which double-blind conditions depend on particular experimental circumstances.

Understanding of the double-blind procedure is further enhanced by empirical analysis and even active experimentation on the kinds and degrees of blindness actually achieved with given techniques. But empirical studies of the double-blind procedure's effectiveness have more than a didactic value; they shed light on the validity of an experiment's assumptions, thereby setting the experimental findings in truer perspective.

That ideal double-blind conditions are often imperfectly fulfilled has, of course, been noted by critics both of the double-blind procedure and of rigorously controlled experimentation in general. But much of this criticism has been unduly nihilistic: since there are circumstances under which desired double-blind conditions cannot be realized, even approximately, the procedure has been labelled a worthless pretense. Yet the empirical findings suggest only that the double-blind procedure is of uneven value, being useful under some experimental circumstances but not others. Clinical as well as laboratory studies are certainly capable of being conducted in complete or near-complete blindness, even with drugs that are potent, provided that circumstances are suitable (as when drug effects are not readily appreciated nor identified by Ss, and when examiners wishing to conduct themselves objectively adopt procedures designed to help them implement this desire).

Even though significant features of an experiment are likely to be transparent to examiners or to Ss, it may yet be useful to adopt double-blind procedures and to adhere to these procedures to the best of one's ability. A biased experiment is, of course, ambiguous in meaning, but as long as an experiment is worth undertaking, there is no point in increasing this ambiguity by encouraging unlimited bias.

The double-blind procedure has both potentialities and limitations, and it is important to explore fully its limits of usefulness. It is especially recommended that investigations employing double-blind procedures include sub-studies designed to evaluate the effectiveness of these procedures by determining the extent to which ideal double-blind conditions are actually attained.

APPENDIX B

THE ARITHMETIC ERRORS TEST

Form A (Pre-treatment)

- A. $5 + 4 = 7 - 2$
- B. $8 \times 4 = 2 \div 1$
- C. $7 - 5 = 6 - 3$
- D. $3 + 4 = 7 + 5$
- E. $6 \div 2 = 4 \times 3$
- F. $6 \div 2 = 8 + 5$
- G. $7 - 4 = 9 \div 6$
- H. $5 + 4 = 3 \div 3$
- I. $8 + 4 = 6 \div 3$
- J. $2 \times 4 = 9 - 3$
- K. $9 - 4 = 4 - 1$
- L. $6 \times 2 = 5 + 3$
- M. $16 \times 4 = 2 \times 2$
- N. $4 - 3 = 8 + 7$
- O. $9 - 3 = 3 \times 1$
- P. $8 \times 2 = 4 \div 4$

Form B (Treatment)

- A. $4 + 5 = 6 - 3$
- B. $8 \times 2 = 4 \div 1$
- C. $5 - 3 = 8 - 4$
- D. $3 + 5 = 8 + 7$
- E. $6 \div 3 = 9 \times 2$
- F. $9 \div 3 = 7 + 4$
- G. $4 - 2 = 8 \div 6$
- H. $7 + 2 = 3 \div 3$
- I. $6 + 2 = 9 \div 3$
- J. $3 \times 2 = 9 - 4$
- K. $8 - 5 = 2 - 1$
- L. $5 \times 3 = 6 + 2$
- M. $12 \times 3 = 2 \times 2$
- N. $6 - 5 = 3 + 2$
- O. $8 - 4 = 2 \times 1$
- P. $9 \times 4 = 6 \div 6$

APPENDIX C

CLUSTERS FOR THE SENTENCE COMPLETION TEST[a]

The Reaction to Stress Cluster:
  3. If they decide to put me under pressure, I . . .
  4. When she thought that the odds were against her, Janet . . .
  9. When they turned him down for the job, Bill . . .
  15. When they said that it was dangerous, I . . .
  18. When Dick failed the course, he . . .
  21. When I meet with difficulties in working out a problem, I . . .
  25. The fact that Pat failed . . .
  32. Discouragement makes me . . .
  38. When luck turned against me, I . . .
  42. When they told me that the job may be too much for me, I . . .
  47. Evelyn's defeat made her . . .

The Reaction to Separation Cluster:
  12. When he left her, Martha . . .
  23. Being left alone . . .
  34. When his friends went away, he . . .
  40. Leaving one's parents . . .

The Expression of Feelings Cluster:
  6. When I get excited, I . . .
  27. When I am irritated, I . . .
  36. When I am angry, I . . .

---

[a] Except for Word Count, results were analyzed only for items included in the three clusters presented here.

REFERENCES

Adler, H. F., Burkhardt, W. L., Ivy, A. C., and Atkinson, A. J.: Effect of various drugs on psychomotor performance at ground level and at simulated altitudes of 18,000 feet in a low pressure chamber. *J. Aviat. Med.,* 1950, *21*:221-236.

Allport, G. W.: *Personality: A Psychological Interpretation.* New York, Holt, 1937.

Allport, G. W., and Vernon, P. E.: *Studies in Expressive Movement.* New York, Macmillan, 1933.

Anderson, R. L., and Bancroft, T. A.: *Statistical Theory in Research.* New York, McGraw-Hill, 1952.

Armsen, P.: A new form of table for significance tests in a 2 × 2 contingency table. *Biometrika,* 1955, *42*:494-511.

Axelrod, J., and Reichenthal, J.: The fate of caffeine in man and a method for its estimation in biological material. *J. Pharmacol. & Exper. Therap.,* 1953, *107*:519-523.

Barmack, J. E.: The effect of benzedrine sulfate (benzyl methyl carbinamine) upon the report of boredom and other factors. *J. Psychol.,* 1938, *5*:125-133.

Barmack, J. E.: The time of administration and some effects of 2 grs. of alkaloid caffeine. *J. Exper. Psychol.,* 1940, *27*:690-698.

Bartlett, M. S.: The use of transformations. *Biometrics,* 1947, *3*:39-52.

Bennett, G. K., Seashore, H. G., and Wesman, A. G.: *Differential Aptitude Tests. Manual.* (2nd ed.) New York, Psychol. Corp., 1952.

Berger, R. M., Guilford, J. P., and Christensen, P. R.: A factor-analytic study of planning abilities. *Psychol. Monogr.,* 1957, *71*: No. 6 (Whole No. 435).

Bradford, J. M., Davis, C. C., Davis, F. B., Derrick, C., Neville, H. R., Spaulding, G., and Willis, M.: The Cooperative Reading Comprehension Tests. Information concerning their construction, interpretation, and use. Princeton, N. J.: Cooperative Test Division, Educational Testing Service. No date.

Brownlee, K. A.: *Statistical Theory and Methodology in Science and Engineering.* New York, Wiley, 1960.
Cardall, A. J.: *Preliminary Manual for the Test of Practical Judgment.* Chicago, Science Research Associates, 1942.
Carlson, A. J., Kleitman, N., Muehlberger, C. W., McLean, F. C., Gulliksen, H., and Carlson, R. B.: *Studies on the Possible Intoxicating Action of 3.2 Per Cent Beer.* Chicago, University of Chicago Press, 1934.
Cochran, W. G., and Cox, G. M.: *Experimental Designs.* (2nd ed.) New York, Wiley, 1957.
Cohen, M. R., and Nagel, E.: *An Introduction to Logic and Scientific Method.* New York, Harcourt, Brace, 1934.
Conway, E. J.: *Microdiffusion Analysis.* London, Crosby Lockwood, 1939.
Cooperative English Test, Single Booklet Edition: Percentile Ranks for High School and College Students. Princeton, N. J.: Cooperative Test Division, Educational Testing Service. No date.
Davis, P. A., Gibbs, F. A., Davis, H., Jetter, W. W., and Trowbridge, L. S.: The effects of alcohol upon the electroencephalogram (brain waves). *Quart. J. Stud. Alcohol,* 1941, *1*:626-637.
Dodge, R., and Benedict, F. G.: *Psychological Effects of Alcohol.* Publ. Carnegie Inst. 1915, No. 232.
Downey, J. E.: *The Will-temperament and Its Testing.* Yonkers, N. Y., World Book, 1923.
Drew, G. C., Colquhoun, W. P., and Long, H. A.: Effect of small doses of alcohol on a skill resembling driving. *Brit. M. J.,* 1958, No. 5103, 993-999.
DuBois, E. F.: *Basal Metabolism in Health and Disease.* (3rd ed.) Philadelphia, Lea and Febiger, 1936.
Dunlap, K.: Medical studies in aviation. IV. Psychologic observations and methods. *J.A.M.A.,* 1918, *71*:1392-1393.
Dunnett, C. W.: A multiple comparison procedure for comparing several treatments with a control. *J. Am. Statis. A.,* 1955, *50*:1096-1121.
Eisenberg, P.: Expressive movements related to feeling of dominance. *Arch. Psychol.,* 1937, No. 211.
Finney, D. J.: *Experimental Design and Its Statistical Basis.* Chicago, University of Chicago Press, 1955.
Fisher, R. A.: *The Design of Experiments.* (4th ed.) Edinburgh, Oliver and Boyd, 1947.
Flory, C. D., and Gilbert, J.: The effects of benzedrine sulfate and caffeine citrate on the efficiency of college students. *J. Appl. Psychol.,* 1943, *27*:121-134.
Freedman, L. Z.: "Truth" drugs. *Scient. Am.,* 1960, *202*:145-154.
Gilliland, A. R., and Nelson, D.: The effects of coffee on certain mental and physiological functions. *J. Gen. Psychol.,* 1939, *21*:339-348.
Goldberg, L.: Quantitative studies on alcohol tolerance in man. *Acta physiol. scandinav.,* 1943, 5: suppl. XVI.
Goldstein, K.: *The Organism.* New York, American Book, 1939.

References

Goodman, L. S., and Gilman, A.: *The Pharmacological Basis of Therapeutics: A Textbook of Pharmacology, Toxicology, and Therapeutics for Physicians and Medical Students.* (2nd ed.) New York, Macmillan, 1955.

Goodnow, R. E., Beecher, H. K., Brazier, M. A. B., Mosteller, F., and Tagiuri, R.: Physiological performance following a hypnotic dose of a barbiturate. *J. Pharmacol. & Exper. Therap.*, 1951, *102*:55-61.

Guilford, J. P. (Ed.): *Printed Classification Tests.* AAF Aviat. Psychol. Program Res. Rep., No. 5. Washington, Govt. Printing Off., 1947.

Guilford, J. P., and Christensen, P. R.: A factor-analytic study of verbal fluency. *Rep. Psychol. Lab.*, No. 17. Los Angeles: University of Southern California, 1956.

Guilford, J. P., Frick, J. W., Christensen, P. R., and Merrifield, P. R.: A factor-analytic study of flexibility in thinking. *Rep. Psychol. Lab.*, No. 18. Los Angeles, University of Southern California, 1957.

Guilford, J. P., Wilson, R. C., and Christensen, P. R.: A factor-analytic study of creative thinking. II. Administration of tests and analysis of results. *Rep. Psychol. Lab.*, No. 8. Los Angeles, University of Southern California, 1952.

Guilford, J. P., Wilson, R. C., Christensen, P. R., and Lewis, D. J.: A factor-analytic study of creative thinking. I. Hypotheses and description of tests. *Rep. Psychol. Lab.*, No. 4. Los Angeles: University of Southern California, 1951.

Hartocollis, P., and Johnson, D. M.: Differential effects of alcohol on verbal fluency. *Quart. J. Stud. Alcohol*, 1956, *17*:183-189.

Hauty, G. T., and Payne, R. B.: Mitigation of work decrement. *J. Exper. Psychol.*, 1955, *49*:60-67.

Hollingworth, H. L.: The influence of caffeine on mental and motor efficiency. *Arch. Psychol.*, 1912, *3*:1-166.

Hollingworth, H. L.: The influence of alcohol. (Part II). *J. Abnorm. Psychol.*, 1924, *18*:311-333.

Hull, C. L.: The influence of caffeine and other factors on certain phenomena of rote learning. *J. Gen. Psychol.*, 1935, *13*:249-274.

Hunt, J. M., and Cofer, C. C.: Psychological deficit. Ch. 32, in Hunt, J. M. (Ed.) *Personality and the Behavior Disorders*, Vol. II. New York, Ronald, 1944. Pp. 971-1032.

Jellinek, E. M., and McFarland, R. A.: Analysis of psychological experiments on the effects of alcohol. *Quart. J. Stud. Alcohol*, 1940, *1*:272-371.

Jones, L. V., and Thurstone, L. L.: The psychophysics of semantics: an experimental investigation. *J. Appl. Psychol.*, 1955, 39:31-36.

Kent-Jones, D. W., and Taylor, G.: Determination of alcohol in blood and urine. *Analyst*, 1954, *79*:121-136.

Kettner, N. W., Guilford, J. P., and Christensen, P. R.: A factor-analytic study across the domains of reasoning, creativity and evaluation. *Psychol. Monogr.*, 1959, *73*: No. 9 (Whole No. 479).

Kraepelin, E.: *Über die Beeinflussing einfacher psychischer Vorgänge durch einige Arzneimittel.* Jena, Gustav Fischer, 1892.

Kruskal, W. H., and Wallis, W. A.: Use of ranks in one-criterion variance analysis. *J. Am. Statis. A.*, 1952, 47:583-621.

Landis, C.: Physiological and psychological effects of the use of coffee. Ch. 4, in Hoch, P. H., and Zubin, J. (Eds.) *Problems of Addiction and Habituation.* New York, Grune and Stratton, 1958. Pp. 37-48.

Lasagna, L.: Placebos. *Scient. Am.*, 1955, 193:68-71.

Lehmann, H. E., and Csank, J.: Differential screening of phrenotropic agents in man: psychophysiologic test data. *J. Clin. Exper. Psychopath.*, 1957, 18:222-235.

Lewin, K., Dembo, T., Festinger, L., and Sears, P. S.: Level of aspiration. Ch. 10, in Hunt, J. M. (Ed.), *Personality and the Behavior Disorders,* Vol. I. New York, Ronald, 1944. Pp. 333-378.

Loomis, T. A., and West, T. C.: The influence of alcohol on automobile driving ability: an experimental study for the evaluation of certain medicolegal aspects. *Quart. J. Stud. Alcohol,* 1958, 19:30-46.

McFarland, R. A.: The psycho-physiological effects of reduced oxygen pressure. Ch. VI, in *The Inter-relationship of Mind and Body. Res. Publ. A. Nerv. & Ment. Dis.*, 1939, 19:112-143.

McFarland, R. A.: *Human Factors in Air Transportation.* New York, McGraw-Hill, 1953.

McFarland, R. A., and Barach, A. L.: The relationship between alcoholic intoxication and anoxemia. *Am. J. M. Sc.*, 1936, 192:186-198.

Marquis, D. G., Kelly, E. L., Miller, J. G., Gerard, R. W., and Rapoport, A.: Experimental studies of behavioral effects of meprobamate on normal subjects, in *Meprobamate and other agents used in mental disturbances. Ann. New York Acad. Sc.*, 1957, 67: Art. 10, 701-711.

Masserman, J. H.: *Principles of Dynamic Psychiatry.* Philadelphia, Saunders, 1946.

Maxwell, A. W.: *Experimental Design in Psychology and the Medical Sciences.* London, Methuen, 1958.

Mead, L. C.: The effects of alcohol on two performances of different intellectual complexity. *J. Gen. Psychol.*, 1939, 21:3-23.

Mellanby, E.: Alcohol, its absorption into and disappearance from the blood under different conditions. *Brit. Med. Res. Council, Special Rep. Ser.*, No. 31, 1919.

Memoirs of the National Academy of Sciences, Vol. 15. Washington, U. S. Govt. Printing Off., 1921.

Miles, W. R.: *Alcohol and Human efficiency; Experiments with Moderate Quantities and Dilute Solutions of Ethyl Alcohol on Human Subjects. Publ. Carnegie Inst.*, 1924, No. 333.

Mirsky, I. A., Piker, P., Rosenbaum, M., and Lederer, H.: "Adaptation" of the central nervous system to varying concentrations of alcohol in the blood. *Quart. J. Stud. Alcohol,* 1941, 2:35-45.

Mullin, F. J., and Luckhardt, A. B.: The effect of alcohol on cutaneous tactile and pain sensitivity. *Am. J. Physiol.*, 1934, *109*:77-78.

Murphy, G.: *Personality: A Biosocial Approach to Origins and Structure.* New York, Harper, 1947.

Nash, H.: Incomplete sentences test in personality research. *Educ. Psychol. Measmt.*, 1958, *18*:569-581.

Nash, H.: The design and conduct of experiments on the psychological effects of drugs. *J. Nerv. & Ment. Dis.*, 1959, *128*:129-147. (Reprinted with modifications in Uhr, L., and Miller, J. G. [Eds.]: *Drugs and behavior.* New York, Wiley, 1960. Pp. 128-155.)

Nash, H.: The double-blind procedure. Rationale and empirical evaluation. *J. Nerv. & Ment. Dis.*, in press.

Neimark, E., and Saltzman, I. J.: Intentional and incidental learning with different rates of stimulus-presentation. *Am. J. Psychol.*, 1953, *66*:618-621.

Newman, H. W.: *Acute Alcoholic Intoxication.* Stanford University, Stanford University Press, 1941.

Newman, H. W.: Experimental approach to the correlation of blood alcohol concentration and intoxication. In *Proceedings of the Second International Conference on Alcohol and Road Traffic.* Toronto, Garden City Press Co-operative, 1955. Pp. 146-150.

Newman, H., and Fletcher, E.: The effect of alcohol on vision. *Am. J. M. Sc.*, 1941, *202*:723-731.

Nowlis, V., and Nowlis, H. H.: The description and analysis of mood. *Ann. New York Acad. Sc.*, 1956, *65*:345-355.

Osol, A., and Farrar, G. E., Jr. (Eds.): *The Dispensatory of the United States of America.* (24th ed.) Philadelphia, Lippincott, 1950. Vols. 1 and 2.

Payne, R. B., and Hauty, G. T.: The effects of experimentally induced attitudes upon task proficiency. *J. Exper. Psychol.*, 1954, *47*:267-273.

Payne, R. B., and Hauty, G. T.: Some psychological factors governing the effects of cerebral depressants upon learned behavior. *Arch. Int. Pharmacodyn.* 1957, *111*:470-477.

Payne, R. E., Moore, E. W., and Bethurum, J. L.: The effects of certain motion sickness preventatives upon psychological efficiency. *USAF Sch. Aviat. Med. Proj. Rep.*, 1952, Proj. No. 21-32-019, Rep. No. 1.

Pemberton, C. L.: A study of the speed and flexibility of closure factors. Unpublished doctoral dissertation, University of Chicago, 1951.

Rosenzweig, S., Fleming, E. E., and Clarke, H. J.: Revised scoring manual for the Rosenzweig Picture-Frustration Study. *J. Psychol.*, 1947, *24*:165-208.

Sapoznik, H. I., Arens, R. A., Meyer, J., and Necheles, H.: The effect of oil of peppermint on the emptying time of the stomach. *J.A.M.A.*, 1935, *104*:1792-1794.

Seashore, R. H., and Ivy, A. C.: The effects of analeptic drugs in relieving fatigue. *Psychol. Monogr.*, 1953, *67*: No. 15 (Whole No. 365).

Sherman, I. C.: A study of Kraepelin's continuous-subtraction test. *J. Abnorm. & Social Psychol.*, 1924, *18*:385-388.

Smith, K. U., Harris, S., and Shideman, F. E.: Effects of ethyl alcohol on the component movements of human motions. *Fed. Proc.*, 1957, *16*:336-337.

Snedecor, G. W.: *Statistical Methods*. (5th ed.) Ames, Iowa State College Press, 1956.

Stanley, W. C., and Schlosberg, H.: The psychophysiological effects of tea. *J. Psychol.*, 1953, *36*:435-448.

Stein, M. I.: The use of a sentence completion test for the diagnosis of personality. *J. Clin. Psychol.*, 1947, *3*:47-56.

Steinberg, H.: Selective effects of an anaesthetic drug on cognitive behavior. *Quart. J. Exper. Psychol.*, 1954, *6*:170-180.

Stone, C. P., Girdner, J., and Albrecht, R.: An alternate form of the Wechsler Memory Scale. *J. Psychol.*, 1946, *22*:199-206.

Street, R. F.: *A Gestalt Completion Test*. New York, Teachers College, Columbia University, 1931.

Strongin, E. I., and Winsor, A. L.: The antagonistic action of coffee and alcohol. *J. Abnorm. & Social Psychol.*, 1935, *30*:301-313.

Stroop, J. R.: Studies in interference in serial verbal reactions. *J. Exper. Psychol.*, 1935, *18*:643-661.

Suhr, W.: Evaluation of work decrement indicators. *Iowa Acad. Sc., Proc.*, 1954, *61*:433-438.

Takala, M., Siro, E., and Toivainen, Y.: Intellectual functions and dexterity during hangover: experiments after intoxication with brandy and with beer. *Quart. J. Stud. Alcohol*, 1958, *19*:1-29.

Terman, L. M., and Merrill, M. A.: *Measuring Intelligence*. Boston, Houghton Mifflin, 1937.

Thornton, G. R., Holck, H. G. O., and Smith, E. L.: The effect of benzedrine and caffeine upon performance in certain psychomotor tasks. *J. Abnorm. & Social Psychol.*, 1939, *34*:96-113.

Thurstone, L. L.: *Primary Mental Abilities*. Chicago, University of Chicago Press, 1938.

Thurstone, L. L.: A factorial study of perception. *Psychometr. Monogr.*, 1944, No. 4.

Thurstone, L. L.: *Mechanical Aptitude III. Analysis of Group Tests*. Chicago, University of Chicago Press, 1949. (Psychometr. Lab. Rep. No. 5.)

Travis, L. E., and Dorsey, J. M.: Effect of alcohol on the patellar tendon reflex time. *Arch. Neurol. Psychiat.*, 1929, *21*:613-624.

Tyler, D. B.: The effect of amphetamine sulfate and some barbiturates on the fatigue produced by prolonged wakefulness. *Am. J. Physiol.*, 1947, *150*:253-262.

Vernon, H. M., Sullivan, W. C., Greenwood, M., and Dreyer, N. B.: The influence of alcohol on manual work and neuromuscular coordination. *Med. Res. Comm. Special Rep. Ser.*, 1919, No. 34.

References

Vogel, M.: Low blood alcohol concentration and psychological adjustment as factors in psychomotor performance; an exploratory study. *Quart. J. Stud. Alchol,* 1958, *19*:573-589.

Von Felsinger, J. M., Lasagna, L., and Beecher, H. K.: The persistence of mental impairment following a hypnotic dose of a barbiturate. *J. Pharmacol. & Exper. Therap.,* 1953, *109*:284-291.

Von Felsinger, J. M., Lasagna, L., and Beecher, H. K.: Drug-induced mood changes in man. 2. Personality and reactions to drugs. *J.A.M.A.,* 1955, *157*:1113-1119.

Watson, G., and Glaser, E. M.: *Watson-Glaser Critical Thinking Appraisal. Manual.* Yonkers, World Book, 1952.

Wechsler, D.: *The Measurement of Adult Intelligence.* (3rd ed.) Baltimore, Williams and Wilkins, 1944.

Wechsler, D.: A Standardized Memory Scale for Clinical Use. *J. Psychol.,* 1945, *19*:87-95.

Wechsler, D.: *The Wechsler-Bellevue Intelligence Scale, Form II. Manual.* New York, Psychol. Corp., 1946.

Wechsler, D.: *Manual for the Wechsler Adult Intelligence Scale.* New York, Psychol. Corp., 1955.

Wells, F. L.: *Mental Tests in Clinical Practice.* Yonkers, World Book, 1927.

Wolff, H. G., Hardy, J. D., and Goodell, H.: Measurement of the effect on the pain threshold of acetylsalicylic acid, acetanilid, acetophenetidin, aminopyrine, ethyl alcohol, trichlorethylene, a barbiturate, quinine, ergotamine tartrate and caffeine: An analysis of their relation to the pain experience. *J. Clin. Invest.,* 1941, *20*:63-80.

Zubin, J.: Objective studies of disordered persons. Ch. 20, in Andrews, T. G. *Methods of Psychology.* New York, Wiley, 1948. Pp. 595-623.

AUTHOR INDEX

A

Adler, H. F., 49, 65, 69
Albrecht, R., 17, 20, 23
Allport, G. W., 50, 52
Anderson, R. L., 38
Arens, R. A., 104
Armsen, P., 142, 145
Atkinson, A. J., 49, 65, 69
Axelrod, J., 36

B

Bacon, F., 146
Bancroft, T. A., 38
Barach, A. L., 62, 70
Barmack, J. E., 40, 42, 43, 63, 106, 107, 120
Bartlett, M. S., 39
Becker, W., 66
Beecher, H. K., 57, 116
Benedict, F. G., 66, 87
Bennett, G. K., 18, 23, 24
Berger, R. M., 18, 19, 20, 24
Bethurum, J. L., 115, 117
Bradford, J. M., 23
Brazier, M. A. B., 116
Brownlee, K. A., 38
Burkhardt, W. L., 49, 65, 69
Bush, A., 66

C

Cardall, A. J., 26
Carlson, A. J., 62, 89

Carlson, R. B., 62, 89
Christensen, P. R., 18, 19, 20, 24, 61, 78
Clarke, H. J., 26
Cochran, W. G., 12
Cofer, C. C., 120
Cohen, M. R., 146
Colquhoun, W. P., 98, 110, 111, 119
Colson, Z. W., 66
Conway, E. J., 15
Cox, G. M., 12
Crozier, W. J., 111
Csank, J., 49, 65, 69, 71, 80

D

Davis, C. C., 23
Davis, F. B., 23
Davis, H., 71
Davis, P. A., 71
Dembo, T., 22
Derrick, C., 23
Dodge, R., 66, 87
Dorsey, J. M., 49, 112
Downey, J. E., 17, 26
Drew, G. C., 98, 110, 111, 119
Dreyer, N. B., 111
DuBois, E. F., 23
Dunlap, K., 119
Dunnett, C. W., 34, 38

E

Eisenberg, P., 52
Exner, S., 54, 55
Farrar, G. E., Jr., 104, 113

F

Festinger, L., 22
Finney, D. J., 12
Fisher, R. A., 12
Fleming, E. E., 26
Fletcher, E., 66, 67
Flory, C. D., 49, 65, 76
Freedman, L. Z., 57
Frick, J. W., 19, 20, 61

G

Gerard, R. W., 66
Gibbs, F. A., 71
Gilbert, J., 49, 65, 76
Gilliland, A. R., 49, 62, 80
Gilman, A., 57, 96, 97, 113
Girdner, J., 17, 20, 23
Glaser, E. M., 20, 25, 26
Goldberg, L., 54, 55, 56, 65, 66, 67, 68, 79, 87, 97, 98, 110, 111
Goldstein, K., 53
Goodell, H., 89
Goodman, L. S., 57, 96, 97, 113
Goodnow, R. E., 116
Graf, O., 110
Granit, R., 111
Greenwood, M., 111
Grether, W. F., 116
Guilford, J. P., 18, 19, 20, 24, 61, 77, 78
Gulliksen, H., 62, 89

H

Halperin, M. H., 111
Hardy, J. D., 89
Harper, P., 111
Harris, S., 54
Hartocollis, P., 57
Hauty, G. T., 43, 118
Hildebrandt, H., 89, 92
Holck, H. G. O., 49, 69
Hollingworth, H. L., 6, 49, 50, 62, 65, 69, 70
Horace, 57
Hull, C. L., 49, 71, 80
Hunt, J. M., 120

I

Ivy, A. C., 49, 65, 69, 120

J

Jackson, J. H., 113, 129
Jellinek, E. M., 50, 66, 67, 71, 74, 76, 82, 114, 116, 117, 118
Jetter, W. W., 71
Johnson, D. M., 57
Jones, L. V., 28

K

Kelly, E. L., 66
Kent-Jones, D. W., 15
Kettner, N. W., 19, 20, 61, 78
Kleitman, N., 62, 89
Kraepelin, E., 5, 6, 119
Kruskal, W. H., 39

L

Landis, C., 97
Lasagna, L., 57, 116, 136
Lederer, H., 97
Lehmann, H. E., 49, 65, 69, 71, 80
Lewin, K., 22
Lewis, D. J., 19
Long, H. A., 98, 110, 111, 119
Loomis, T. A., 57, 70, 83
Luckhardt, A. B., 87

M

McDougall, W., 116
McFarland, R. A., 50, 62, 66, 67, 70, 71, 74, 76, 82, 111-119 passim
McLean, F. C., 62, 89
Marquis, D. G., 66
Masserman, J. H., 109
Maxwell, A. W., 12
Mead, L. C., 76, 140
Mellanby, E., 14
Merrifield, P. R., 19, 20, 61
Merrill, M. A., 20
Meyer, J., 104
Miles, W. R., 50, 55, 66, 89
Mill, J. S., 146
Miller, J. G., 66
Mirsky, I. A., 97
Moeller, H., 66
Moore, E. W., 115, 117
Mosteller, F., 116
Muehlberger, C. W., 62, 89
Mullin, F. J., 87
Murphy, G., 52

Author Index

N

Nagel, E., 146
Nash, H., 10, 25, 60, 63, 91, 104, 107, 134, 146, 147
Necheles, H., 104
Neimark, E., 27
Nelson, D., 49, 62, 80
Neville, H. R., 23
Newman, H. W., 66, 67, 97, 108
Nowlis, H. H., 10
Nowlis, V., 10

O

Osol, A., 104, 113

P

Payne, R. B., 43, 115, 117, 118
Pemberton, C. L., 24
Piker, P., 97
Powell, W. H., 66

R

Rapoport, A., 66
Reichenthal, J., 36
Roget, P., 57
Rosenbaum, M., 97
Rosenzweig, S., 26

S

Saltzman, I. J., 27
Sapoznik, H. I., 104
Schlosberg, H., 49, 69
Sears, P. S., 22
Seashore, H. G., 18, 23, 24
Seashore, R. H., 49, 120
Sherman, I. C., 23
Shideman, F. E., 54
Siro, E., 66
Smith, E. L., 49, 69
Smith, K. U., 52, 54
Snedecor, G. W., 36, 39
Spaulding, G., 23

Stanley, W. C., 49, 69
Stein, M. I., 25
Steinberg, H., 116
Stone, C. P., 17, 20, 23
Street, R. F., 21
Strongin, E. I., 54
Stroop, J. R., 22
Suhr, W., 49, 69
Sullivan, W. C., 111

T

Tagiuri, R., 116
Takala, M., 66
Taylor, G., 15
Terman, L. M., 20
Thornton, G. R., 49, 69
Thurstone, L. L., 21, 23, 24, 28, 64
Toivainen, Y., 66
Travis, L. E., 49, 112
Trowbridge, L. S., 71
Tyler, D. B., 120

V

Vernon, H. M., 111
Vernon, P. E., 52
Vogel, M., 98
Von Felsinger, J. M., 57, 116

W

Wallis, W. A., 39
Watson, G., 20, 25, 26
Wechsler, D., 17, 20, 21, 23, 25, 80
Wells, F. L., 18
Wesman, A. G., 18, 23, 24
West, T. C., 57, 70, 83
Willis, M., 23
Wilson, R. C., 19
Winsor, A. L., 54
Wolff, H. G., 89
Wundt, W., 5

Z

Zubin, J., 119

SUBJECT INDEX

For entries devoted to tests, descriptions of procedure are indicated by *italicized* page references, and the principal presentations and discussions of drug findings are indicated by **boldface** page references.

A

Abstract Reasoning, *18*, **76-77**, 118, 121
Acuity
 auditory, 74
 visual, 66-68, 74, 76
Addition
 continuous adding, 63, 106-107
 practical implications of findings, 109
 present task, *19*, **58**, **62-63**, 112, 114, 120, 121
Age, 9, 33, 34
Aggressiveness, 25, 26, 88, 90-91
Alcohol dosage level
 design of present study, 6-7, 12-15, 95, 99, 110
 regression on, 57-59, 95-99, 110-114, 128-129
Amphetamines
 Barmack's findings, 63, 106
 versus caffeine, hand steadiness effects, 126
 versus methyl-caffeine, 43
 research, historical aspects, 6
Analysis of variance
 versus blood level analysis, 95-99

 heterogeneity of variance, 38, 39
 procedure, *36-39*
 tabulated analyses, 42, 51, 61, 65, 69
Anoxia, 115-116, 119
Arithmetic errors, *19*, 37, **78-79**, 82, 120, *149*
Associate learning, *20*, 37, **71-77** passim, 118
Associations, 37, **56-63**
 biphasic effects, 59, 60, 62, 109, 112-114, 117
 constrained, 56-57, 61-63
 disrupted, 89-90, 92
 neutralized gains hypothesis, 80
 quality, 58, 61, 62-63, 114
 summary, 126, 127
 unconstrained, 56-60
 word, 56-59
Attention
 boundary condition for measuring capacity, 63, 107
 and detection of drug effects, 53, 55, 118-119
 drug effects on, 42, 44
 findings mediated by drug effects on, 63, 70, 106-107

163

Attitude
and Barmack's hypothesis, 40-43, 63, 106-107
and blood-taking procedure, 10, 87, 89
communication of, 10, 135, 146
and double-blind procedure, 133-148
self-ratings, 40-44
and writing expansiveness, 52
Auditory acuity, 74
Automobile driving, 70, 84, 108-109, 110, 111

B

Barbiturates
and association, 57, 113
and task complexity, 116
Barmack's hypothesis, 40-43, 63, 106-107
Biphasic effects, 128-129
on association, 59, 60, 62, 109, 112-114
Blocks
effects, 36
examiner's awareness of, 142
interactions, 38
pre-drug testing, 10
randomized, defined, 12
Blood alcohol level, present study
beginning- vs. end-of-session values, 52, 84
design of study, 13-15, 95, 99, 110
values obtained, 13, 35-36
Blood alcohol level, regression on
Goldberg's study, 55-56, 65-68, 79, 97-98, 110-111
present study, 36, 95-99
yet other studies, 57, 66, 98, 110-111
Blood alcohol level, standard of intoxication, 108-109
Blood caffeine level, 13, 14, 15, 35-36
Blood-taking procedure, 9, 10, 11, 15, 95
See also Puncture, reaction to
Body surface area
determination of, 23
differences between treatment groups, 33, 34
dosage dependent on, 13-14
and double-blind procedure, 139
Boredom, 42-43, 63, 106-107

C

Capacity
versus attitude, 63, 106-107, 128
definition, 45
over-all functioning, 7, 45, 94, 103-106, 125-126
Cerebellar functioning, 54-56, 119
Cerebrospinal axis, descending inhibition, 6, 113-114, 115, 128-129
Clerical speed and accuracy, 21, **64-65**, 68, 70, 120, 121
Closure 1, 64-65, 67-68
Coffee, 34, 62, 109
Color naming, 22-23, 56, 61-62
Compensation for impairment
judgment, 84, 108, 128
writing expansiveness, 53, 118-119, 127
Composite measures
components with equal scale factors, 37, 40, 58, 64, 68-69, 71, 76
components with unequal scale factors, 37
versus independent measures, 46, 47
procedure, 37, 38-39
Confounded drug effects
attitude and capacity, 63, 106-107, 118-119, 128
fatigue and capacity, 49, 52-53, 63, 106-107
neutralized gains, 80
practice effects, 69, 79, 105-106, 118
side effects, 104
suggestion and biased drug findings, 105, 127, 133-139, 146
Consequences, *19*, **91-92**
active efforts at mastery—passivity, 88, **92-94**
favorable—unfavorable outcome, 88, **92-94**
low-quality, 37, **56-59**, **61**
remote, 37, **58**, **61**
Constructive response, under stress, 37, 88, 90

Subject Index

Contingencies, *24*, **81**, **82**, **83**, **85**
Continuous subtraction, *23*, **78**, **79**
Controlled associations, *20*, 37, **58**, 59
Correlations
    pre- versus post-drug, 120
    reaction to stress (projective), 90, 92-94
    with vocabulary, 34, 60, 93
    with writing expansiveness, 52, 53, 54
    writing intensity, inter-rater reliability, 18, 21
Creativity and alcohol, 61, 77, 80, 109
Customary consumption
    alcohol, 33, 34, 67, 96-99
    caffeine, 33, 34, 97

D

Deduction, *20*, **76-77**
Delayed recall, 72-75
Deliberately adaptive task features, 48, 50, 53, 58, 118-119
Depressants, c.n.s., 5-6, 57, 113-114, 115-122, 127-129
Design, experimental
    and double-blind procedure, 133-148
    of Barmack's study, 106-107
    of this study, 6-8, 9-15, 95, 99, 110, 139-141
    See also Sessions; Statistical procedure; Treatments
Digit symbol, *21-22*, **68-69**, **68-70**, 75-76, 77, 120, 121, 127
    attitude towards one's performance on, 71, 72, 81
Digits backward, *17-18*, **75**, **78**, **79-80**, 116-117, 118
Digits forward, 17, 37, **71-73**, **75**, **79-80**, 116-117
Disinhibition hypothesis, 49, 109, 112-117 passim, 128-129
Dispensatory, 113
Doses, in present study, 11-15, 139-140

E

Energy
    augmentation of energy reserves, 80

energy/structure hypothesis, 56-57, 62
    mobilization of aggressive energies, 25, 88, 90-94
Estimation of slow writing, corrected, 27, 81
Estimation of treatment effects, 27, **81**, **83**, 144
Evaluation of arguments, *25-26*, 37, **84-85**
Examiner blindness
    actual extent of, 104, 141-143, **145-148**
    means of safeguarding, 12, 15, 138-139, 140-141
    need for, 134-135
Examiner's guess, 11, 141-143
Experimental error
    computation of error terms, 38
    factors affecting, 12, 44, 47-48, 134, 136
    instances of excessive, 50, 79, 82-83, 85, 86, 90
    See also Length, test; Pre-drug values lacking; Reliability, test
Expressive measures, 45, 50, 53, 118-119

F

Fatigue
    and continuous performance, 49, 63, 106-107
    and digit symbol, 75
    self-ratings, 40-43, 52, 63, 144
    and writing expansiveness, 52-53
    and writing speed, 49
Flexibility, 77
    adaptive, 37, 78, 79
    creativity, 109
    harmonizing distinct activities, 70
    manipulating ideas, 77-81
    suppressing habitual response, 89-90
Free association, *20-21*, 37, **56-59**, 120, 121

G

Gottschaldt figures, *22*, 37, **64-65**, 78
Grape juice, 14, 140

H

H-test
  procedure, 39
  tabulated results evaluated by, 72, 81, 87-88
Habituation, *see* Customary consumption
Hand steadiness, 49, 126
Hypermetria, 55
Hypotheses
  Barmack's, 40-43, 63, 106-107
  disinhibition, 49, 109, 112-117 passim, 128-129
  energy/structure, 56-57, 62
  neutralized gains, 80
  See also Biphasic effects; Recall hypotheses; Task features, hypotheses; Writing expansiveness

I

Immediate, intentional recall, 37, 70-75, 79-80, 121-122, 126-127
Incidental learning, 11, 26-27, **71-72**, 74
Incidental recall, 71-75
Incidental task features, 50, 53, 55, 56, 60, 118-119
Initial dose, 11, 13, 14
Interaction
  among subjects, 10, 137, 140
  among tests, 10
Interaction, drugs and functions, *see* Selective drug effects
Interaction, drugs x sub-plot factors
  results, non-significant, 49, 75-76, 80
  results, significant, 40, 51, 61, 64, 70, 75-76
  statistical procedure, 38, 39
  tabulated analyses, 42, 51, 61, 65, 69

J

Judgment, critical, 81-84, 108, 115
  quality of associations, 62-63, 114
  self-ratings, 41, 44, 83-84
  See also Practical Judgment

L

Language usage, 24, 81, 121

Learning, 75-76, 106
  practice effects, 69, 79, 105-106, 118
  task familiarity, 117-118
  See also Recall
Length, test
  brief tests, 49, 79, 80, 82, 86
  prolonged tests, 42-43, 63, 106-107
  summary, drug effects as a function of, 119-121
Logarithmic transformations
  in Goldberg's study, 55-56, 65, 67, 97, 110-111
  procedure, 39
  tabulated findings, 37, 48, 58, 81, 87

M

Maintenance dose, 11, 13-15, 35-36
Match problems, *18*, 37, **77-78**
Mathematical ability, 115
  adding, 19, 62-63, 106-107, 109, 114
  correcting faulty equations, 19, 78-79, 149
  subtracting, 23, 78, 79
Medico-legal standards, intoxication, 108-109
Memory
  organization, 74-75, 122
  retention, 71-74
  See also Delayed, Incidental, and Immediate, intentional recall
Methyl-caffeine, 43
Missing data, procedure, 12, 29, 36, 38, 39, 104
Motion sickness preventatives, 117, 118
Motor function
  capacity, 48-50, 95-96, 112
  digit symbol, 66-70
  eye co-ordination, movement, 66-67
  incidental aspects, 50-56, 96, 98, 99, 118-119
  prolonged tasks, 112, 119
  reflexes, 49-50, 112
  steadiness, 49, 126
  tracking, 54-55, 110, 111
  See also Automobile driving; Puncture, reaction to
Mutilated Words, 24, 37, **64, 65**, 120, 121

Subject Index

N
Neurosis and alcohol, 109
Nitrous oxide, 57, 116

O
Over-all functioning, 7, 45, 94, 103-106, 125-126

P
Pain, 87, 89
Peppermint oil, 14, 104, 140
Perceptual speed, 37, 64-65, 67-68
Picture Completion, 24-25, 37, 84-85
Placebo, in other studies
   baseline for evaluating results, 63, 69
   composition, 57, 140
   review, 146
Placebo, in present study
   baseline for evaluating results, 34, 38-39, 105-106, 139-140, 142
   composition, 12-14, 139-140, 143-144
   efficacy of double-blind procedure, 105, 141-146
   experimental design, 10, 12-15, 139-140
Placebo and double-blind procedure, general considerations, 133, 136-138, 147
Plot titles, 20, 37, 58, 59, 61
Post-battery procedures, 11, 26-28
   See also Sessions
Power tests, 24, 77, 85, 121-122
Practical judgment, 26, 37
   understanding how to handle an emergency, 81, 84, 92
   understanding the reasons for accepted social practices, 85
Practice, 69, 79, 105-106, 118
Pre-drug values lacking
   marked effects despite, 60
   measures insensitive because, 47, 50, 82, 85, 86, 90
   and over-all capacity findings, 46, 47
   test procedures, 16, 24-29
Projective techniques, *see* Consequences; Rosenzweig Picture-Frustration; Sentence Completion

Psychiatric disturbance
   alleviation by drugs, 5, 109
   indices of, 79-80, 119
Puncture, reaction to, 11, 28-29, 90, 92
   dislike puncture, 41, 44, 87, 89
   tightening, withdrawal, 15, 37, 86-87, 89

R
Randomization
   randomized blocks design, 12-13
   of test batteries, 10, 11, 35, 82-83
   of treatments, 12
Rapid functioning
   and examination atmosphere, 63, 107, 135, 146
   in power tests, 77, 85, 121-122
   in speeded tests, 17, 63, 77, 121-122
Rating scales
   categories, 18-29 passim, 43
   of motor response, 15, 18, 21, 29
   polarity, 43
   reliability, 18, 21, 43, 44
   See also Puncture, reaction to; Self-ratings
Recall hypotheses
   auditory acuity, 74
   organization, 74-75, 122
   retention, 71-74, 122
Regression
   on alcohol dosage level, 6-7, 57, 95-99, 110-114, 128-129
   on caffeine dosage level, 7, 49
   on customary drug consumption, 67, 96-99
   See also Blood alcohol level, regression on
Reliability, inter-rater, 18, 21
Reliability, test
   findings obscured by low, 43-44, 50, 51, 67, 78, 80, 85
   inclusion of exploratory tests, 8, 16
   summary, 119-121
   See also Length, test
Rosenzweig Picture-Frustration, 26, 37, 88-94 passim

S
Selective drug effects, 7

association, 56-57, 61, 62
motor control, 49, 53, 54, 126
rapid functioning, 121-122
recall, 74, 80
summary, 126-127, 128-129
See also Interaction, drugs x sub-plot factors
Self-ratings, 11, 28, 40-44
boredom, 42-43, 63
fatigue, 40-43, 52, 63
judgment, 44, 83-84, 144-145
stress, response to, 44, 86-89
Sentence completion, 25, 60, *150*
expression of feelings, 86, 87
reaction to separation, 37, 88
reaction to stress, 37, 88-94 passim
word count, 58, 59-61, 114, 119
Sentence memory, *18*, 37, **72**, 73
Sessions, 10-12, 16-29
differences between, 10, 11, 40, 41
variations within, 11, 35, 52, 82-83, 84
Side effects, alcohol, 35, 103-105, 143
Sign, negative
and enhancement, 46, 65, 89
and impairment, 45
Slow writing, *see* Writing, slow
Speeded tests, 17-25, 60, 63, 77, 121-122
Statistical procedure
analysis of variance, 36-39
comparison of treatments, 38-39, 40, 41
non-parametric, 39
transformation of data, 37, 39
See also Correlations; Missing data, procedure; Randomization; Regression; Statistical significance
Statistical significance
of capacity findings, 46, 47
and Dunnett allowance, 38-39, 41
versus practical significance, 107-109
two-tailed tests, 38-39, 103
and validity of findings, 93
Stimulants, c.n.s., 5, 6, 63, 106-107, 113, 120, 127-129
Story memory, 23-24, 37, **72-75**
Street Gestalt completion, *21*, 37, **64, 65**, 67, 120

Stress, *performance* under, *see* Puncture, reaction to; Stroop Color-Word
Stroop Color-Word, *22-23*, 46, 77, 86, 92
color-naming Time, 58, **61-62**
conflict time, 87, 89
reading time, 63, 64, **65**
Stroop ratio x 100, 87, **89**
Subject blindness
actual extent of, 104, 143-146, 147-148
means of safeguarding, 10, 135-138, 139-140
need for, 133, 135
Subject sample
characteristics, 33-35
compensation by intellectual Ss, 84, 108
and double-blind procedure, 135-138, 140
selection criteria, 9
Subjects not completing experiment
and confounded results, 104
and heavy alcohol dose results, 35, 82, 83, 99, 143, 144, 145
illness of, 35
Sub-plot factors, *see* Interaction, drugs x sub-plot factors
Suggestion, 105, 134-135
See also Examiner blindness; Subject blindness

T

Task features, hypotheses, 115-122
See also Incidental task features; Length, test; Rapid functioning
Test
alternate forms, 16-29 passim, 105, 120, 149
order of presentation, 10, 11, 13-15, 16-29
procedure, 10-29
reliability, inter-rater, 18, 21
task features, 115-122
See also Pre-drug values lacking; Power tests; Reliability, test; Sessions; Speeded tests; Trials

Subject Index

Time estimation, 27-28, 81
Tired and weary self-ratings, 28, 40-43, 52-53
Tracking, 54-55, 110, 111
Treatments
  doses, 11-15, 139-140
  evaluation of, procedure, 36-39
  judging their effects, 81, 83-84, 141-146
  as primary independent variable, 95, 99, 110
  See also Placebo
Trials
  digit symbol, 21-22, 68-70, 75
  and learning, 75-76
  rote memory tests, 17, 20, 75
  self-ratings, 28, 40-42
  writing expansiveness, 17, 21, 51-53, 95-96, 98-99
  Writing Speed, 17, 21, 49, 95-96
Two-tailed tests
  Dunnett allowance, 38-39
  versus one-tailed tests, 103

U

Understanding, *see* Practical Judgment

V

Validity
  of projective findings, 92-94
  and vocabulary score, 34, 60, 93
Validity, caffeine findings
  addition, 63, 107
  self-ratings, 42-43
Validity, heavy alcohol dose findings
  and side effects, 104
  and suggestion, 105, 146
Visual capacities, 37, 62-77 passim, 81, 108
  See also Acuity; Closure 1; Flexibility; Motor function; Perceptual speed
Vocabulary, 11, 23, 33, 34
  drug findings and vocabulary level, 34, 60, 93

W

Weight, body, *see* Body surface area
Whiskey estimate, 27, 81, 83, 143-144
Word count, *see* Sentence completion
Word fluency, 21, 37, 56-59, 121
Would Trust My Judgment self-ratings, 28, 41, 44, 83-84
Writing expansiveness, 17, 21, 48, 51, 119, 120
  cerebellar hypothesis, 54-56, 119
  fatigue hypothesis, 52-53
  incidental task features hypothesis, 50, 53, 55, 56, 60, 119
  mood hypothesis, 52
  regression on blood alcohol level, 36, 96, 98, 99
Writing intensity, 18, 21, 37, 48, 50-51, 54
Writing, slow, 26, 27, 31, 48, 50
  estimation of, corrected, 27, 81
Writing speed, 11, 17, 21, 48, 49-50
  regression on blood alcohol level, 36, 95-96
  and writing expansiveness, 53, 118-119

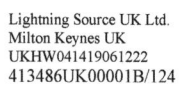